INDOOR AND OUTDOOR AIR POLLUTION

Edited by **José A. Orosa**

Indoor and Outdoor Air Pollution
Edited by José A. Orosa

Published by InTech
Janeza Trdine 9, 51000 Rijeka, Croatia

Publishing Process Manager Natalia Reinic

Technical Editor Teodora Smiljanic

Cover Designer InTech Design Team

Additional hard copies can be obtained from orders@intechopen.com

Indoor and Outdoor Air Pollution, Edited by José A. Orosa
p. cm.
ISBN 978-953-307-310-1

For EU product safety concerns:
IN TECH d.o.o., Prolaz Marije Krucifikse Kozulić 3, 51000 Rijeka, Croatia,
info@intechopen.com or visit our website at intechopen.com.

Meet the editor

José Antonio Orosa graduated in Marine Engineer and Naval Architecture at the University of A Coruña. His Ph.D. has proved the possibilities of energy saving with permeable coverings. Today, he is Head of the Department of Energy of the University of A Coruña. During the last years he has participated in the International Energy Agency Annex 41 and collaborated with the INEGA and IDEMEC of the University of Porto in research about indoor air quality.

Contents

Preface

The atmosphere is the layer of air surrounding the earth and its structure below 50 km is most important in determining pollution levels. In this sense, greenhouse effect and climate change are evoking a special interest recently since they are considered to be human hazards. Greenhouse effect is produced by infrared radiations, imprisoned between the earth and the thin layer of greenhouse gases, which get reflected and heat up the earth's surface. These gases allow the sun's radiations of shorter wavelengths to pass through when they are reflected from the surface of the earth, but do not allow the radiations of longer wavelengths, which are the hot infrared radiations, reflected from the earth to go out because of which there is an increase in temperature on the surface of the earth. The temperature of the earth increases by almost half a degree per century and this phenomenon tends to rapidly increase causing climate change and influencing human health.

Greenhouse effect and climate change are not the only human hazards to be considered. Indoor and outdoor air quality must also be considered in this context. Air quality is a measure of the degree of ambient atmospheric pollution. Deterioration and damage to both public health and environment due to poor air quality have been recognized at a legislative and international level. Air pollution is often quantified for purposes of comparison or threshold attainment using the air quality index (AQI). This and other indexes will be shown in-depth in Chapter 1.

On the other hand, the ambient concentration of criteria pollutants found within the atmosphere of a domain can be predicted or simulated by numerical simulation. The principal application of air pollution modeling is to investigate the air quality scenario so that the associated environmental impact on a selected area can be predicted and quantified. The Geographical Information System (GIS) was applied to this type of analysis in order to organize the data results. In accordance with this point of view and previous concepts, different case studies about outdoor and indoor air quality have been analyzed in this book.

The first case study, Chapter 2, shows how to model the particulate matter released during and after the application of biosolids based on the data collected during the field study. The objective of this case study was to develop and evaluate a dispersion model for particulate matter associated with biosolid application on a farm field. The efforts include a derivation of solution to a convective–diffusion equation incorporating wind shear.

Air pollutants are continuously released from numerous sources into the atmosphere. Several studies have been carried out on the quantification of pollutants and their consequences on public health. Identification of the source characteristics of air pollution is an important step in the development of regional air quality control strategies. In this sense, the aim of the second case study, Chapter 3, was to analyze the environmental data gathered on the daily monitoring of ambient ozone, oxides of nitrogen, and sulfur (IV) oxide at five monitoring sites. Furthermore, in recent years, certain statistical techniques that incorporate the influence of meteorological variables have been applied to assess the trend in ozone levels in ambient air. One common approach is the use of a parametric regression model to link different parameters to meteorological variables. In this approach, another of the case studies shows, in Chapter 4, how the cooper materials can be corroded by Santa Ana Winds.

After the conduct of case studies on outdoor air, it is time to comment on indoor air. In this book, different topics on indoor air quality are discussed. One case study shows, in Chapter 5, a correlation of professional performance to acceptable IAQ in critical care medical facilities. This case study is especially interesting with regard to, at present, the quality of the indoor environment for the health and safety of occupants as well as the protection of assets, which is becoming a prime concern with end users. Other case studies conducted a review of the economic valuation studies as a theoretical framework of welfare economics and provided a brief description of the fundamental issues arising in such environmental problems, as we can see in Chapter 6. It is an interesting point of view and the currently available methods for estimating monetary values of changes in well-being due to air pollution are shown.

To summarize, this book tries to reveal different points of view of the wide concept of air quality. In this context, there will be an initial introductory chapter on the main concepts of air quality, following which there will be real case studies on outdoor and indoor air quality with an aim to provide a guideline for future standards and research works.

José A. Orosa
Departamento de Energía y P.M.
Universidade da Coruña
Coruña,
Spain

Part 1

Outdoor Air Quality

1

Air Polluted Environment and Health Effects

Michael Theophanides, Jane Anastassopoulou and
Theophile Theophanides
*National Technical University of Athens, Chemical Engineering School,
Radiation Chemistry & Biospectroscopy,
Greece*

1. Introduction

1.1 The general problem of pollution

The natural environment in which we live in is ever-increasingly threatened by human activity (Theophanides, M. et al 2002). Both the inhabited and uninhabited environment is threatened and one such indication is the changes of the climate (Theophanides, T. et al. 2002). Furthermore, as of 2006, the International Union for Conservation of Nature and Natural Resources (*IUCN) Red List* contains over 15,000 species threatened with extinction (M. Theophanides et al. 2007, 2007, Touloumi et al. 1994, Katsouyanni 2003, Arribas-Monzón, et al. 2001, 1998, Yang, et al. 2004; Kotzias, 2003). The assessment includes species from a broad range of taxonomic groups including vertebrates, invertebrates, plants, and fungi. Human health is threatened with diseases and early mortality and is even more prevalent in emerging economies facing rapid industrialization. There is increasing evidence that global warming also contributes to a higher rate of disease growth and propagation. Epidemiological studies (The MACBETH project 1999: *IT070*, Jerrett M. et al 2004; Samoli E.et al 2003, Tunnicliffe et al. 2003; Filleul L, et al 2003, Basu R. & Samet J. M., 2002, Le Tertre A. et al. 2002, Dominici F., 2002, Sunyer & Basagana 2001) of occupational diseases on the working population are showing the ill effects of the environment on people working in a contaminated environment over a lifetime of employment. The study of occupational diseases is becoming an ever-increasing problem to be investigated (Kunzli 2001). The social and economic (Kunzli 2001) evolution inevitably produced stress situations in the environment resulting in population density increases that were difficult to handle efficiently using existing infrastructures and continuing increasing urbanization of cities. On the other hand, while we are quite aware of the sources of pollution, a great deal of research is still needed to recognize the effects on health, the climate, extinguishing species and their role in the evolutionary and food-supply chain. We must also educate individual citizens about pollution - starting with even the simplest of things such as recycling, reducing their dependency on the automobile, and not littering.

Exposure to pollution from gaseous pollutants diminishes the quality of atmospheric air that we breathe can induce diseases and deaths in increased numbers in the population when the values of pollutants are exceeding the recommended safe thresholds. The most vulnerable to such effects are the elderly, children and those already afflicted with health problems.

Several studies (M. Theophanides et al. 2007, 2007, Touloumi et al. 1994, Katsouyanni 2003, Katsouyanni et al. 1997, Ballester, et al 1996, Arribas-Monzón, et al. 2001) indicated a positive association of sulfur dioxide (SO_2) and nitrogen oxide (NO_2) with mortality - NOx being one of the principal emissions of aviation industry. Benzene is a well-known carcinogen (Touloumi et al. 1994) depending on the degree of exposure, and can affect persons with indoor or outdoor air exposure for which the risk of death can be higher. A number of studies in recent years substantiate the detrimental effect of environmental pollution on human health, disease and pollution (Theophanides, M. 2002 , 2007; Touloumi et al. 1994, Katsouyanni 2003, Katsouyanni et al. 1997, Ballester, et al 1996, Arribas-Monzón, et al. 2001).

1.2 What is air quality?

"Air Quality" is a measure of the degree of ambient atmospheric pollution, relative to the potential to inflict harm on the environment. The potential for deterioration and damage to both public health and the environment, through poor air quality, has been recognized at a legislative and international level. Air pollution is often quantified for purposes of comparison or threshold attainment using the Air Quality Index (AQI).

The Air Quality Index (AQI) has been developed by the Environmental Protection Agency (EPA) USA, to provide accurate, information about daily levels of air pollution. The Index provides organizations with a standardized system of measuring pollution levels for the major air pollutants that are regulated. Index figures enable the public to determine whether air pollution levels in a particular location are Good, Moderate, Unhealthy for Sensitive Groups or worse. In addition, EPA and local officials use the AQI as a public information tool to advise the public about the general health effects associated with different pollution levels and to describe whatever precautionary steps may need to be taken if air pollution levels rise into the unhealthy range.

The EPA uses the Air Quality Index to measure five major pollutants for which it has established National Ambient Air Quality Standards under the Clean Air Act (Tobias et al. 2001). The pollutants are particulate matter, sulfur dioxide, carbon monoxide, nitrogen dioxide and ground level ozone. For each of the five pollutants, EPA has established air quality standards protecting against health effects that can occur within short periods of time (a few hours or a day). For example, the standard for sulfur dioxide - that is, the allowable concentration of this pollutant in a community's air - is 0.14 parts per million measured over a 24-hour period. Air concentrations higher than 0.14 parts per million (ppm) exceed the national standard. For ozone, the 8-hour average concentration permitted under the standard is 0.085 parts per million (ppm).

In the USA, the AQI is calculated every hour for each air quality parameter according to the following formula (Coull, 2001):

$$AQI = \frac{I_{HI} - I_{LO}}{BP_{HI} - BP_{LO}} X(C_{03} - BP_{LO}) + I_{LO}$$

AQI=Air Quality Index, I_{LO}=Index at the lower limit of the AQI category, I_{HI}= Index at the upper limit of the AQI category

BP_{LO}= Break-point concentration at lower limit of the AQI category, BP_{HI}= Break-point concentration at upper limit of the AQI category, C_{O3}=8-hour ozone concentration

Parameter	Concentration	Units	AQI Formula
Carbon Monoxide	If > 13	ppm	AQI = (1.47 x concentration) + 5.88
	If <= 13		AQI = 1.92 x concentration
Ozone	If <= .05	ppm	AQI = 500 x concentration
	If > .05 <= .08		AQI = (833 x concentration) - 16.67
	If > .08		AQI = (714 x concentration) - 7.14
Sulfur Dioxide	All	ppm	AQI = 147.06 x concentration
Nitrogen Dioxide	If <= 0.21	ppm	AQI = 238.09 x concentration
	If > 0.21		AQI = (156.24 x concentration) + 17.19
PM2.5	If <= 30	ug/m^3	AQI = 0.8333 x concentration
	If > 30		AQI = (0.5 x concentration) + 10

Table 1. Air Quality Index Formula

Table 2 illustrates the likely health effects from various levels of AQI based on the US standards:

AQI Range	EPA Color Scale	EPA Descriptor	Clean Air Campaign Health Advisory
0 to 50	Green	Good	The air quality is good and you can engage in outdoor physical activity without health concerns.
51 to 100	Yellow	Moderate	At this level the air is probably safe for most people. However, some people are unusually sensitive and react to ozone in this range, especially at the higher levels (in the 80s and 90s). People with heart and lung diseases such as asthma, and children, are especially susceptible. People in these categories, or people who develop symptoms when they exercise at "yellow" ozone levels, should consider avoiding prolonged outdoor exertion during the late afternoon or early evening when the ozone is at its highest.
101 to 150	Orange	Unhealthy for Sensitive Groups	In this range the outdoor air is more likely to be unhealthy for more people. Children, people who are sensitive to ozone, and people with heart or lung disease should limit prolonged outdoor exertion during the afternoon or early evening when ozone levels are highest.
151 to 200	Red	Unhealthy	In this range even more people will be affected by ozone. Most people should restrict their outdoor exertion to morning or late evening hours when the ozone is low, to avoid high ozone exposures.
201 to 300	Purple	Very Unhealthy	Increasingly more people will be affected by ozone. Most people should restrict their outdoor exertion to morning or late evening hours when the ozone is low, to avoid high ozone exposures.
Over 300	Black	Hazardous	Everyone should avoid all outdoor exertion.

Table 2. Air Quality Index threshold levels (EPA)

A simplified version of AQI is shown in Table 1(Coull, 2001). The highest number calculated for a specific hour is used as the AQI for that hour and indices range from 0 to 100%. Calculating the general equation for specific pollutants results in the pollutant AQI shown in Table 1.

The AQI places maximum emphasis on acute health effects occurring over very short time periods - 24 hours or less - rather than chronic effects occurring over months or years. By notifying the public when an AQI value exceeds 100, citizens are given an adequate opportunity to react and take whatever steps they can to avoid exposure. The approach EPA follows is conservative, because (1) each standard has built into it a margin of safety that is designed to protect (1) highly susceptible people, and (2) the public notice is triggered as soon as a single sampling station in the community records an AQI level that exceeds 100.

Finally, the AQI does not take into account the possible adverse effects associated with combinations of pollutants (synergism). As more research is completed in the future, the AQI may be modified by EPA to include such effects.

1.3 What is air pollution numerical simulation?

Numerical Simulation of Air Pollution is the attempt to predict or simulate, by numerical means, the ambient concentration of criteria pollutants found within the atmosphere of a domain. The principal application of air pollution modeling is to investigate air quality scenarios so that the associated environmental impact on a selected area can be predicted and quantified. It is important in several ways (Coull 2001).

i. To aid in the evaluation of source-receptor relationships so that responsibility for specific impacts can be apportioned.
ii. To aid in project planning, site evaluation and/or environmental impact of present/future sources.
iii. To enable the evaluation of existing sources in relation to compliance with legislation.
iv. To permit the evaluation of proposed abatement and control strategies, in relation to short and/or long term issues.
v. To permit the assessment of episodic tactics and disaster aversion strategies
vi. To optimize emission inventories and operating conditions while ensuring compliance with legislative controls.

It was possible to compare the results of simulation data with the specific air data that had been calculated in an area and correlates with pollution levels of the region. The Geographical Information Systems (GIS) was applied to this type of analysis in order to organize data results. This study integrates atmospheric simulation chemical data collected in various forms and emissions data into a GIS environment. In the study gas samples were collected and added to the GIS database. High resolution GIS models were created for a few regions where various types of atmospheric simulation studies were conducted. The dispersion of combined pollutants NO_x, VOCs, Benzene, PM is shown in Fig. 1.

The dark red corresponds to higher levels of pollutants and indicates the dispersion along the industries and agricultural lands from the point of pollutant sources. The dispersion direction depends on atmospheric conditions.

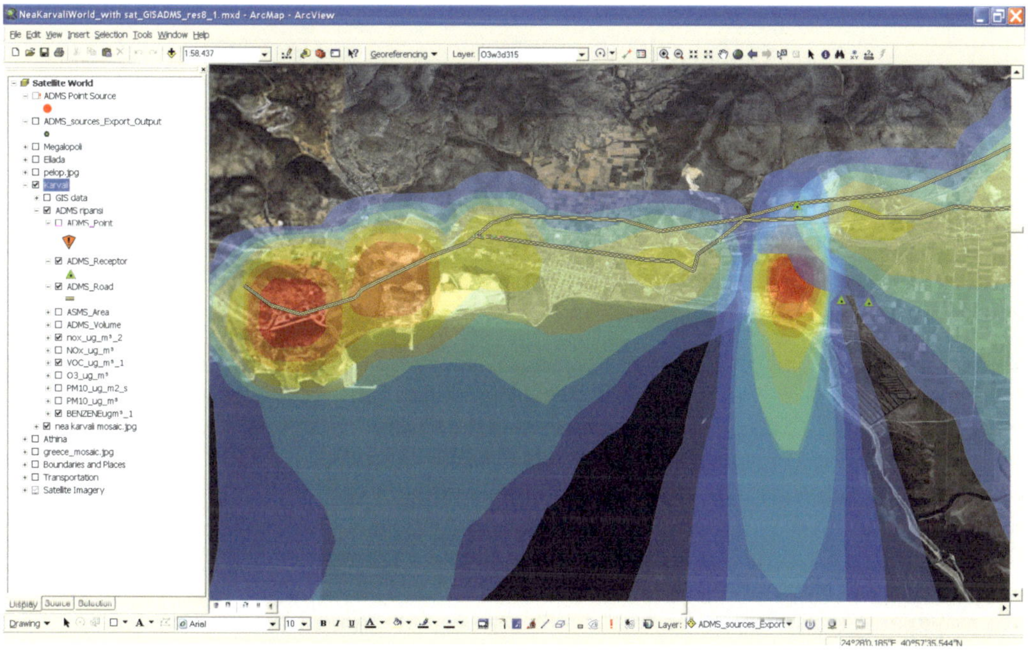

Fig. 1. Numerical simulation of pollution dispersion including all factors, in Kavala Greece

1.4 Measurement units

The measurement of trace concentrations of gases can be expressed in several different ways in literature. *Parts per million* (**ppm**) can be expressed *by volume* or *by mass* which is the main source of confusion. For example, if a pie is divided into 1 million pieces, then 1 ppm is 1 piece of the pie ($1x10^{-6}$). In this case, being a solid, it is ppm by mass. Sometimes **ppmv** is used to remind us that it is by volume. By volume (e.g. gases), the molecular weight must be considered:

$$1ppm \frac{V_m x1\mu g\ gas}{M1litre\ air}$$

V_m = 22.711 litres/mol = standard molar volume of ideal gas at 1 bar, 273.15 °K,
M = molecular weight of gas
Therefore, comparing grouped pollutants such as VOCs, HC and PM expressed in ppm is not always appropriate since they are made up of many compounds that have varying molecular weights. *Parts per billion* (ppb), is similar in concept and is $1x10^{-9}$ (1 nanogram) per cubic meter, ng/m^3.
The other important unit is $\mu g/m^3$. It simply expresses, with no ambiguity, the quantity of gas present in a given volume. From the point of view of pollutants as health hazards, ppm is a less relevant measure since equal portion of pollutants (expressed in ppm) do not result in the same health hazard. Table 3 shows the conversion from one unit to the other.

Gas	Description	Molecular weight	V_m/M	(5ppm) µg/m³
CH_4	Methane	16	1.4194425	3.53
H_20	Water Vapour	18	1.2617267	3.96
CO	Carbon monoxide	28	0.8111111	6.17
NO_2	Nitrous dioxide	46	0.4937191	10.13
O_3	Ozone	48	0.4731475	10.57
C_6H_6	Benzene	78	0.2911677	17.17

Table 3. Conversion from 5 ppmv to µg/m³ for different compounds

2. The composition of the atmosphere

The atmosphere is the sphere of air surrounding the earth. The structure of the atmosphere below 50 km (50,000 meters) is most important for pollution considerations (See Fig.2). The troposphere comprises the part of atmosphere from ground level up to 11,000 meters. This section is generally characterized by turbulent weather, low ozone (O_3) levels, high water content (H_2O) and a linearly varying temperature from ISA conditions on ground to -55 C at the limit of its height of 11 km. The atmosphere is relatively dense and approximately 80% of the atmospheric mass is contained in the troposphere. Approximately half of the solar radiation reaches the surface.

The tropopause is marked by the delineation between the troposphere and the stratosphere. The temperature is a constant -55 C and it is at 11 km above the earth's surface (Figure 2). The stratosphere is a region of upper atmosphere stretching from the tropopause (11km) to approximately 50 km from the earth's surface. It is generally characterized by high content of ozone (O_3) and very low water content (H_2O).

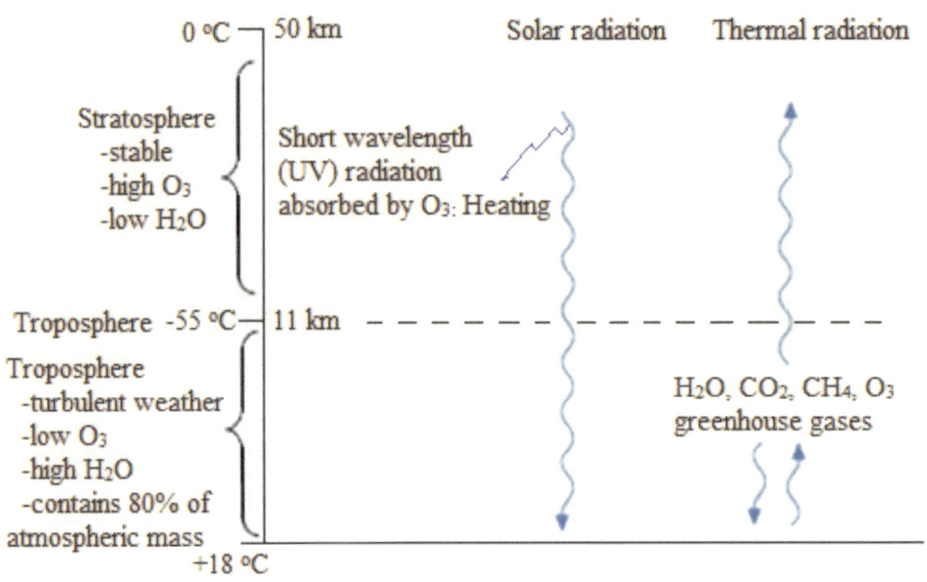

Fig. 2. Structure of the Atmosphere, Radiation and Greenhouse Effect

It is substantially more stable environment with very little vertical mixing. As ultra-violet solar radiation from the sun is absorbed by ozone (O_3) when it passes through the stratosphere, the result is a heating of the upper atmosphere up to a maximum of 0 C at 55 km from the earth's surface.
Ultra-violet (UV) solar radiation is absorbed by ozone (O_3) as it passes through the atmosphere, heating the upper portion of this region and causing a temperature maximum near 50 km. Below this, some of the solar radiation is reflected, mainly by clouds, and some is absorbed but about half gets through to the surface. This heats the near surface region and results in a second temperature maximum, this time at the surface. The tropopause marks the sharp boundary between the troposphere, in which the temperature drops markedly with height, and the stratosphere, where it generally increases with height. Various atmospheric constituents allow most of the short-wave solar radiation through but absorb and then re-emit the long-wave thermal radiation. This warms the near surface region, the so-called greenhouse effect. Water vapor (H_2O), carbon dioxide (CO_2), methane (CH_4) and ozone (O_3) are examples of important "greenhouse gases". A convenient measure of the greenhouse effect of a change in a constituent is provided by the imbalance between solar and thermal radiation at the tropopause when the change in the constituent is suddenly imposed.
At the top of the atmosphere, the solar energy absorbed by the Earth/atmosphere is balanced by the emission of longer wavelength thermal radiation (heat). However, the thermal radiation emitted from the near surface region is absorbed by greenhouse gases, which then re-emit back towards the surface, keeping it warm. The heat lost to space is from levels typically near 5 km where the air is colder than at the surface.

2.1 The fixed gases in the atmosphere

Understanding the natural composition of the earth's atmosphere is necessary to understand the consequences and nature of the substances that are constantly being added to our atmosphere. The main composition of the lower atmosphere is shown in Table 4 and consists mostly of Nitrogen (N_2) and Oxygen (O_2) forming up to 99% of all molecules. The remaining 1 % are trace concentrations of inert gases helium, neon, argon, krypton and xenon and appear in the concentrations specified in Table 4.

Fixed Gas	%	ppmv
Nitrogen (N_2)	78.08	780,000
Oxygen (O_2)	20.95	209,500
Helium (He)	0.0005	5
Neon (Ne)	0.0015	5
Argon (Ar)	0.93	9,300
Krypton (Kr)	0.0001	1
Xenon (Xe)	0.000005	0.05

Table 4. Fixed Gases of the Atmosphere

Fixed gases are well mixed in the atmosphere and have stable mixing ratios. The following Table 4 summarizes the key contributors and they are described individually in greater detail below (Sommer et al. 1999).

Molecular Nitrogen

Molecular Nitrogen is produced biologically in soils. During the growth of bacteria in anaerobic environments nitrate (NO^-_3) is reduced to N_2 and small amounts of nitrous oxide gas (N_2O) in what is known as "denitrification". The source of nitrate in the soil occurs from a two-step 'nitrification' process from ammonium (NH_4^+). Ammonium is produced in three ways:

i. Naturally from the decomposition of organic material which contains nitrogen atoms
ii. Naturally from a process called nitrogen-fixation occurring in aerobic environments whereby some amounts of N_2 are converted to ammonium (NH_4^+)
iii. Man-made generation such as fertilizers and other industrial processes

However, this production process of molecular nitrogen is slower than denitrification and, therefore, the concentration of N_2 has increased in the atmosphere over time.

Molecular Oxygen

Molecular Oxygen is produced by photosynthesis when CO_2 reacts with H_2O in the presence of solar radiation and chlorophyll found in trees, plants, and algae. A product of this process is carbohydrates of the form $C_nH_{2n}O_n$. For example, when n=6, glucose is derived:

$$6CO_2 + 6H_2O + hv \xrightarrow{\text{chlorophyll-l}} C_6H_{12}O_6 + 6O_2$$

2.2 Variable and trace gases in the atmosphere

Variable gases have volume mixing ratios that can change significantly over time and vary according to location. They are anthropogenic in origin. Natural processes or atmospheric pollution due to human activity in many circumstances can directly affect their concentration levels. The following Table 5 summarizes the key elements of the variable gases.

Variable Gas	%	ppmv
Water Vapor (H_2O)	0.00001 – 4.0	0.1 – 40,000
Carbon Dioxide (CO_2)	0.0360	360
Methane (CH_4)	0.00017	1.7
Ozone (O_3)	0.000003 – 0.001	0.03 – 10

Table 5. Variable Gases of the Atmosphere

2.3 Volatile organic compounds and hydrocarbons

Volatile Organic Compounds (VOCs) are organic volatile chemicals that have high vapor pressure and will easily form vapor at standard ambient temperature and pressure. The term is generally applied to organic aromatic compounds such as benzene, toluene, ethylbenzene, m/p-xylene and o-xylene, organic solvents, aerosol spray can propellants, fuels (gasoline, kerosene), petroleum distillates. VOCs are also naturally emitted by a number of plants and trees. Many VOCs are flammable. VOCs can be removed with special filtration systems such as activated charcoal systems that absorb organic materials.

VOCs are an important health and environment concern for several reasons:

i. Some VOCs can be hazardous to health when inhaled. Benzene is a known human carcinogen and toxic. Likewise, Formaldehyde is both an irritant and a sensitizer as well as being toxic
ii. VOCs such as hydrocarbons constitute emissions that instigate photochemical smog.
iii. Some VOCs such as methyl-tertbutyl-ether (MTBE) are gasoline additives that are fairly soluble in water leading to the contamination of water.
iv. VOCs can form particulate matter if condensation forms of the gas

Volatile Organic Compounds, Polycyclic Aromatic Hydrocarbons (PAH) and the carbonyl compounds (CO, CO_2, CH_2O) are the major organic pollutants in the atmosphere. The assessment of VOCs has become a major issue of air quality network monitoring in cities. Toluene and Benzene are the major pollutants. Table 6 gives some examples of common VOCs.

VOC			
Hydrocarbons			**Other**
Aromatic	**Aliphatic**	**Non-aromatic**	
Unsaturated Planar Structure Cyclic Noticeable smell	Saturated Tetrahedral Linear or cyclic	Unsaturated Chain-like structure Non-cyclic	
Examples : Bezene	Hydrogenated products	Omega-3	Formaldehyde Liquor

Table 6. Volatile Organic Compound Classifications

2.4 Particulate matter (PM)

The term particulate matter (PM) is used to describe airborne solid particles such as dust, dirt, soot, smoke and/or liquid droplets. Classification by size is based on the aerodynamic diameter because it is a good indicator of the transport and removal of particles from the air and their deposition within the respiratory system. Based on size, urban PM tends to be divided into three principal groups: *coarse* (PM_{10}: 2.5 µm -10 µm), *fine* ($PM_{2.5}$: 1 µm - 2.5 µm) and *ultra fine* particles ($PM_{0.1}$: < 0.1 µm) as shown in Fig. 3. Coarse PM (PM_{10}) include geological materials, pollen and sea salt to name a few examples.. Fine PM is derived principally from emissions such as SO_2, NO_x and condensation of VOCs. Ultrafine PM ($PM_{0.1}$) including nanoparticles ($PM_{0.01}$) result from condensed organic carbon or sulfuric acid vapors (Fig. 3).

The particles contained in the PM_{10} size fraction may reach the upper part of the airways and lung.

2.5 The green house gases

Water vapor, Carbon dioxide, and Methane are important greenhouse gases. A greenhouse gas is a gas that readily absorbs infrared radiation. Methane, for example, absorbs infrared radiation emitted by the earth 21 times more efficiently than CO_2 (per molecule). However, because mixing ratios are much higher for CO_2, it has a more significant impact on the greenhouse effect.

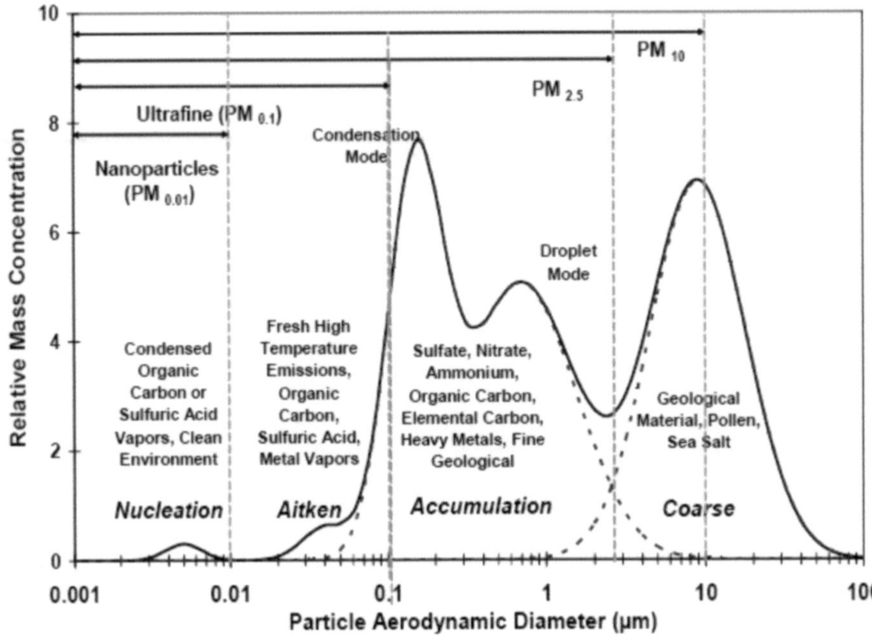

Fig. 3. PM Size chart

It is Jean-Baptiste Fourier who first discovered in 1824 the importance of the greenhouse effect and the overheating of the atmosphere. A little later, the chemist Nicolas-Theodore de Saussure imagined that the emission of CO_2 from immense fires could delay the melting of the glaciers. However, towards the end of the XIX century John Tyndall and Olaf Arrhenius discovered the role of the carbon dioxide and the water vapors in this effect. Since then we have learned that there are other gases than the CO_2, which contribute to the greenhouse effect, like nitrogen dioxide (NO_2) and various chlorofluorohydrocarbons (CFC) produced by the industry, such as CCl_3F (trichlorofluoromethane) called (CFC-11) and Cl_2F_2 (dichlorodifluoromethane) called CFC-12.

The greenhouse effect is produced by the infrared radiations, which are imprisoned between the earth and the thin layer of the greenhouse gases, which are reflected and heat the earth's surface. These gases let go through the sun's radiations of short wavelengths, such as the visible and ultra -violet, when they are reflected from the surface of the planet and keep the radiations of longer wavelengths, which are the hot infrared radiations that increase the temperature of the surface of the earth. The temperature of the Globe increases almost half a degree (0.5 C) per century (from 13.84 C in 1950 to 14.4 C in 2001) and this phenomenon tends to increase rapidly.

3. The impact of the transport industry on human health

3.1 Road transport pollution

It is only very recently that the full extent of transport's negative impact on health has become clearer. In an ecological audit of the impact of cars on German society[52] concluded that cars were responsible for 47,000 deaths each year and a range of other, less severe, health impacts. These are summarised in Table 7.

	Number	Unit
deaths from particulate pollution	25,500	deaths, pa
deaths from lung cancer	8,700	"
deaths, from heart attacks	2000	"
deaths from summer smog	1900	"
deaths from road traffic accidents (RTAs)	8758	"
	TOTAL 47,000	"
serious injuries (RTAs)	116,456	injured/pa
light injuries (RTAs)	376,702	"
chronic bronchitis (adults)	218,000	number of illnesses/pa
Invalidity due to chronic bronchitis	110	number of invalidates/pa
coughs/auswurf	92,400,000	days/year
bronchitis (children)	313,000	number of illnesses/pa
Wiedreholt Husten	1,440,000	number of illnesses/pa
Hospitalisation (breathing problems)	600	number of hospitalisations/pa
Hospitalisation (breathing problems)	9,200	number of days of care/pa
Hospitalisation (cardiovascular disease)	600	hospitalisations/pa
Hospitalisation (cardiovascular disease)	8,200	number of days of care/pa
Arbeitsunfahigkeit (not cancer)	24,600,000	
Asthma attacks (days with attacks)	14,000,000	days/pa
Asthma attacks (days with broncho-dilator)	15,000,000	days/pa

Table 7. Health damage caused by cars, Germany, 1996, annual totals

The volume of death and illness revealed in Table 7 puts the European transport problem into a very serious public health perspective. Transport is a major health problem and should be tackled as much within a public health context as in a traditional transport/roads/highway context. All the deaths and injuries in Table 7 relate to cars and not to lorries or aircraft. Total deaths are about 5 times greater than road traffic accidents deaths.

The total amount of sickness, days in hospital etc. imposes a huge burden on the health services of European countries and this burden is not recovered from those who drive cars. The health impact is a huge human tragedy. 15 million days of use of bronchodilators is a huge problem for many children and many families and the impact on physical activity, social activity, enjoyment of outdoor pursuits, community and neighbourhood is incalculable. Health impacts in Europe in the 21st century are the direct equivalent of disease impacts in 19th century cities which then required major re-engineering with clean drinking water and sewage systems. We are still waiting for the 21st century equivalent of this re-engineering to deal with the modern equivalent of widely dispersed sewage.

Road traffic noise and noise from aircraft also create significant health problems (WHO, 1996). These health problems are generally understated in Europe with an implicit assumption on the part of traffic engineers and planners that most people can get used to noise and, in any case, it is only a minor irritation and part of life in an advanced industrial society. This has to be rejected. Noise causes raised blood pressure, cardiovascular disease, a range of psychological problems, sleep disturbance and it damages school age children if they are exposed to noise in a learning environment. WHO (1993) discusses the evidence that supports the contention that children exposed to noise learn less well and have reading abilities lower than is the case for children not in noisy environments. Studies around Heathrow Airport in SE England also point to damage to children living near the airport and under flight paths.

Studies of individual exposure to pollution show that car occupants are exposed to 2-4 times as much pollution from vehicles as are cyclists[53]. This finding is in some ways counter-intuitive and surprising but is the result of cars following a very similar path through traffic to that followed by all other cars and effectively driving in a "tunnel of pollution". This raises the very interesting and important conclusion that the car itself damages the health of car occupants. The conventional view is that cars are safer and more pleasant than cycling (presumed to be a dangerous activity). Scientific research shows that this is not the case and the growth of car use in Europe (especially the increase in the number of children carried around by car) represents a significant public health problem which is at least an example of direct correspondence between perpetrator and victim. Those that cause the problem suffer the consequences of that problem. The fastest increase occurred between 1950 and 2000, which is due to human activity.

3.2 Aviation pollution

Commercial aviation is experiencing dramatic growth in regions throughout the world, including North America and the United States. However, airport development has not kept pace with increases in aviation activity and the problem is now acute in the United States. In 1996 the Federal Aviation Administration (FAA) Administrator identified lack of airport capacity as the "single most important constraint" to realizing forecast rates of growth throughout the aviation industry. Funding is one problem. The annual shortfall between funds needed for airport development and total funds available is difficult to determine but has been estimated at more than $4 billion annually in recent years. A second problem is the rapid pace of change in aviation technologies. Changes in the design and construction of airfield and landside facilities will be necessary to accommodate the larger aircraft that will enter service and the new navigation and air traffic control systems that will be deployed in the near future.

According to IPCC, in 2000 aviation was responsible for 3% of carbon dioxide emissions due to the total burning of fossil fuel and 13% of that associated with transport. However, the total greenhouse impact was more important than this would suggest. Since the vast majority of the flights were subsonic and therefore in the 9 - 13 km height range, the emissions of oxides of nitrogen led, on average, to an increase in ozone as well as a decrease in methane.

Innovative planning approaches are essential to timely development of new airport facilities, and environmental documentation is a key component of the planning process. Federal actions (e.g., funding, approvals) in connection with proposed airport development often require environmental review pursuant to the National Environmental Policy Act (NEPA) and the implementing guidelines of the Council on Environmental Quality and the FAA, which is in the process of updating its *Airport Environmental Handbook* (Order 5050.4A). In addition to NEPA, a number of states have enacted statutes that mandate evaluation of the potentially significant environmental impacts of development, including airport projects. Beyond compliance with NEPA and state environmental review statutes, airport development proposals may trigger additional analytic requirements that must be carried out in parallel or sequential processes, for example, air quality assessments pursuant to the Environmental Protection Agency (EPA) General Conformity Rule and historic resource documentation pursuant to Section 106 of the National Historic Preservation Act of 1966. Environmental analyses for airport development projects are increasingly subject to technical, political, and ultimately legal scrutiny. More and more often, challenges are raised as to the adequacy of NEPA and state environmental documents as well as studies supporting related determinations by lead agencies or agencies with jurisdiction or special expertise (Health Canada 2005).

The main environmental concerns associated with aircraft are:

- Climate change
- stratospheric ozone reduction, leading to increased surface UV radiation
- regional pollution - changes in tropospheric chemistry for tens to hundreds of kilometres downwind of the airport. In particular, emissions of oxides of nitrogen in air increase ozone
- local pollution - both noise and decreased air quality caused by aircraft and also by the associated ground transportation.

There is no doubt that both local pollution and regional pollution are very serious issues. It is thought that European Directives on permitted levels of oxides of nitrogen may limit the expansion of some airports (Filliger et al 1999). Concern has been expressed to us that the techniques for assessment of the impact of aircraft emissions on both local and regional air quality are poorly developed and that the available modeling tools are in general inadequate. This issue needs addressing urgently, especially in the light of the recent consultation documents on regional airport development. However, the focus here is on the possible larger scale impacts of aviation, on surface UV radiation through changes in atmospheric ozone and on climate.

3.3 Key pollutants

The main types of pollution linked to aviation and airport operations originate from aircraft, ground-support equipment (GSE), external traffic related to airport activity and industrial parks. Aircraft operations are related primarily to pollution such as carbon monoxide and dioxide, nitrogen oxides, oxides of sulfur, water vapor, hydrocarbon trace pollutants such as benzene and particulate matter consisting mainly of sulfate and soot. These emissions alter the chemical composition of the atmosphere in a variety of ways, both directly and indirectly. On the larger-scale, sulfur oxides in aircraft emissions are important only as a source of particles.

The unique feature of these emissions is that the majority of them occur far above the Earth's surface. Subsonic aircraft generally cruise in an altitude range of 9 - 13 km, close to the tropopause, the sharp transition between the troposphere and the stratosphere (see Fig. 1). The troposphere is the region in which the turbulent motions and precipitation related to weather occur. In contrast the stratosphere is relatively stable and the vertical motions in it are generally sufficiently small compared with the horizontal motions that the air travels almost horizontally (Filliger P et al 1999).

3.4 Avition and the atmosphere

The impact of aircraft emissions can be very different depending whether they are in the upper troposphere or the lower stratosphere. Both the abundance of trace gases and the dominant chemical composition and associated chemical reactions are very different in the two regions. In particular water vapour content is relatively high in the troposphere and low in the stratosphere, whereas ozone levels are much higher in the stratosphere. Stratospheric ozone absorbs radiation from the sun. This leads to a heating profile in the stratosphere that determines its character, and also protects life at the surface from the harmful effects of the UV radiation.

The height of the troposphere varies with latitude. In the tropics the tropopause is higher than the normal range of subsonic cruise altitudes but in Polar Regions it is usually at the lower end of this range. Whether an aircraft cruises in the upper troposphere or the lowermost stratosphere depends on its location, the weather and the time of year. Supersonic aircraft typically cruise at levels in the range 17 - 20 km, which is always in the stratosphere (Filliger et al, 1999). Jet streams are typically located at the tropopause in regions where there are abrupt transitions in the horizontal between the troposphere and the stratosphere. Since eastward-flying aircraft are often routed in the strong westerly winds in jet stream regions to save fuel and time, they often fly close to this almost vertical tropopause. The dominant physical and chemical processes differ between the troposphere and stratosphere, as do the time-scales for transporting air between regions. Water vapor added by any human activity in the troposphere is soon lost through mixing and precipitation processes, whereas at 20 km it persists and moves slowly towards the pole.

A "conservative gas" is one that becomes well mixed throughout the atmosphere so that the point of emission is irrelevant for its impact on climate. The carbon dioxide produced by the combustion of kerosene in aircraft engines behaves as a conservative gas and so becomes well mixed. However, oxides of nitrogen, produced by high temperature burning in the engine, are rapidly involved in chemical reactions that lead to changes in both ozone and ambient methane. These reactions are complex and sensitive. Ozone is generally produced by oxides of nitrogen in the troposphere and destroyed by it in the lower stratosphere. Since the lifetime of ozone is relatively short, its aircraft-induced increase or decrease is restricted in both the vertical and the horizontal. The lifetime of methane, however, is sufficiently long that the reduction in it produced by the emitted oxides of nitrogen becomes distributed throughout the atmosphere. In the troposphere the amount of water vapor emitted in aircraft exhaust is negligible compared with the pre-existing concentrations in the atmosphere. However, along with the particles emitted, the water vapor can lead to condensation trails, some of which can persist for hours and perhaps trigger the development of cirrus clouds. Subsequent cirrus cloud may also be further influenced by particles emitted by aircraft (Samet et al. 2000).

4. Exposure to various air pollutants and health effects

4.1 Clean air

Clean air is the symbol of life for humans, animals and vegetation and forms the basis of the food processing mechanisms of all these three life-borne species. Oxygen is the active ingredient in the air that reacts with food supplies by oxidizing them or burns them in various animal and vegetable tissues to maintain the balance of life. The remainder of the fixed gases (nitrogen and inert gases) does not react with the food supply. Damage to the food processing mechanism impacts growth and reproduction and, therefore, the future of life. The variable gases can have a direct effect on the quality of air necessary to maintain the food chain mechanisms. Pollution and toxic chemicals can have a detrimental effect on the balance of life and the food chain.

The chemicals that are poisoning our environment are numerous. They change the composition of the atmospheric air. In addition, there are traces of solid materials ($PM_{0.1\text{-}0.01}$, $PM_{2.5}$, PM_{10}) in the air that are equally toxic - usually metal oxides and other solid compounds. Species that breathe air to burn food with oxygen also inhale the toxic chemicals that interact or react with the animal tissues. Lungs, in particular, are susceptible to such damage leading to symptoms of pulmonary diseases that can range from acute irritations to chronic illnesses or death.

Toxic substances can also work their way into the blood stream and cause cardiovascular diseases. It is also known that chemicals may damage hemoglobin and react with tissues such as breast, lung and heart tissues and cells or constituents of cells such as proteins, nucleic acids, membranes such as lipids, phospholipids, and carbohydrates. These possible chemical alterations in the above organs and molecules can lead to other diseases such as cancers and osteoporosis.

Today, people agree that the treasures of the earth are finite and that the biosphere is vulnerable. Studies indicate that 80% of all materials produced by companies become trash within six months[36]. Until now, our environment has been able to cope with this massive influx of pollution but we are starting to see its limitations of absorbing. The time is rapidly approaching where we must make dramatic changes in the way we manage our environment to avoid drastic consequences to future generations. Increased earth temperature caused by the greenhouse effect will facilitate incubation of bacteria and lead to increases in diseases. Flooding or droughts will also lead to poor hygiene for millions of humans and produce other diseases. As recent catastrophes such as SARS and the poultry viruses demonstrated, humans halfway across the globe in well-developed, western nations are equally susceptible to the transfer of these diseases. Therefore, pollution and environmental management are issues that must be addressed by the developed world.

The results of epidemiological studies can be applied to current air quality statistics to estimate the magnitude of the impact of air pollution on health. The World Health Organization (WHO) produced meta-analyses for the effects on mortality and morbidity of a number of pollutants (WHO, 1997). Their effect estimates have been used by others to calculate aspects of the burden of poor health attributable to pollution. For example, in the

UK, COMEAP (the UK Department of Health's Committee on the Medical Effects of Air Pollutants) calculated that PM10 was associated with 8,100 deaths brought forward and with 10,500 emergency hospital respiratory admissions (brought forward and additional) in urban areas of Great Britain. The corresponding figures for SO_2 were 3,500 deaths brought forward and 3,500 early and extra hospital admissions. The effects of ozone were 700 deaths and 500 admissions if there is no health effect below 50ppb, but 12,500 and 9,900 if there is no threshold. Our own studies have demonstrated that high levels of toxic air pollution can be correlated to increased mortality (Theophanides, M . et al. 2002, 2007).

4.2 Exposure assessments

Several empirical methods have been devised to quantify the effect of pollution on mortality and morbidity. The WHO has Meta-Analysis for the effects on mortality and morbidity of a number of pollutants (WHO, 1997) for PM_{10}, SO_2, O_3. Furthermore, the WHO and is increasingly leaning toward the conclusion, substantiated by supporting research, that for some pollutants, there is no threshold below which is deemed safe. At the very least, increasingly sensitive epidemiological study designs have identified adverse effects from air pollution at increasingly lower levels. For the time being, linear models are being used for which there is no lower threshold (WHO, 2003).

In short-term studies, elderly subjects, and subjects with pre-existing heart and lung disease were found to be more susceptible to effects of ambient PM on mortality and morbidity. In panel studies, asthmatics have also been shown to respond to ambient PM with more symptoms, larger lung function changes and with increased medication use than non-asthmatics. In long-term studies, it has been suggested that socially disadvantaged and poorly educated populations respond more strongly in terms of mortality. PM also is related to reduce lung growth in children. No consistent differences have been found between men and women, and between smokers and non-smokers in PM responses in the cohort studies. (WHO, 2003)

4.3 Quantification of effects

The quantification of health effects has become increasingly important in the development of air quality policy. For such analyses it is important to have accurate information on the concentration–response relationships for the effects investigated, i.e. on the relationship between the level of air pollution and the effect on health. A quantitative meta-analysis of peer-reviewed European studies was therefore conducted to obtain summary estimates for certain air pollutants and health effects. The data for these analyses came from a database of time-series studies developed at St George's Hospital Medical School at the University of London. The meta-analysis was performed at St George's according to a protocol approved in advance by a WHO Task Group. Using data from several European cities, the analysis confirmed statistically significant relationships between mortality and levels of PM and ozone in ambient air. Updated risk coefficients in relation to ambient exposure to PM and ozone were obtained for all-cause and cause-specific mortality and hospital admissions for respiratory and cardiovascular causes. Some results are shown in Fig.4. The meta-analysis also included a thorough assessment of so-called publication bias. Fig. 4 shows that $PM_{2.5}$ presents the high mortality risk for an increase in concentration of this pollutant. The most susceptible cause of mortality is cardiovascular deaths.

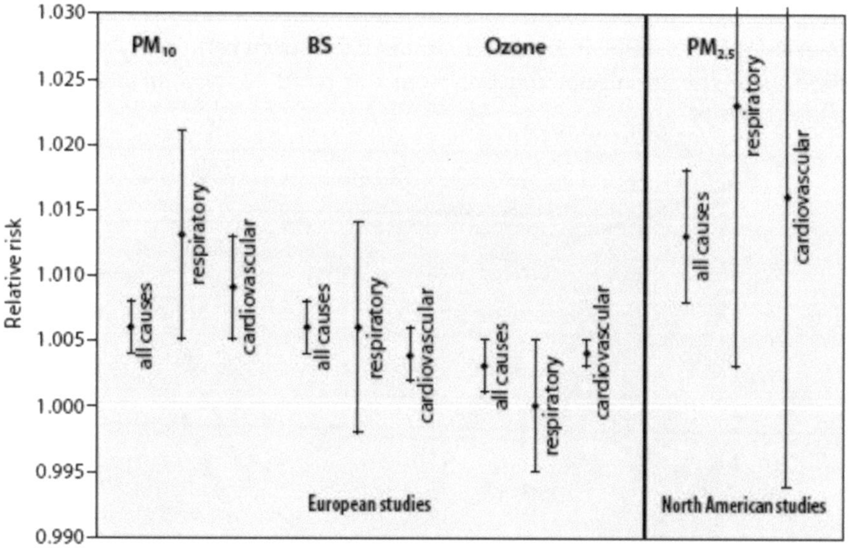

Fig. 4. Probability of mortality risks as a result of a 10 µg/m^3 increase of a pollutant

The WHO has estimated that in Europe 100,000 death are due to air pollution each year. Studies by the Canadian government have concluded that yearly deaths due to air pollution (CO, NO_2, SO_2, PM, O_3) from anthropogenic sources are 1,800 for short-term exposure and 4,200 for long-term exposure (Mcdonnell, w.f. et al. 2000).

4.4 Health effects due to particulate matter (PM)

The effects of short-term exposure to PM have been documented in numerous time-series studies many of them conducted in Europe (Rita Rita K. Seethaler et al 2003); these indicated large numbers of outcomes, such as attributable deaths and hospital admissions for cardiovascular and respiratory conditions. Both short-term (24 hours) and long-term (annual average) guidelines are therefore recommended. The WHO defines the principal short and long-term health effects attributed to Particulate Matter according to Table 8:

Short-Term	*Long-Term*
Lung inflammatory reactions Respiratory symptoms Adverse effects on the: Increase in medication usage Increase in hospital admissions Increase in mortality	Increase in lower respiratory symptoms Reduction in lung function in children Increase in chronic obstructive Reduction in lung function in adults Reduction in life expectancy, owing

Table 8. Health Effects due to PM

Fig. 5 shows schematically where particles are deposited in the respiratory tract, depending on their size. Smaller particles (in particular $PM_{2.5}$) penetrate more deeply into the lung and may reach the alveolar region. Ultrafine particles contribute only slightly to PM_{10} mass but may be important from a health point of view because of the large numbers and high surface area. They are produced in large numbers by combustion (especially internal combustion) engines. As stated above, PM in ambient air has various sources. In targeting control measures, it would be important to know if PM from certain sources or of a certain composition gave rise to special concern from the point of view of health, for example owing to high toxicity.

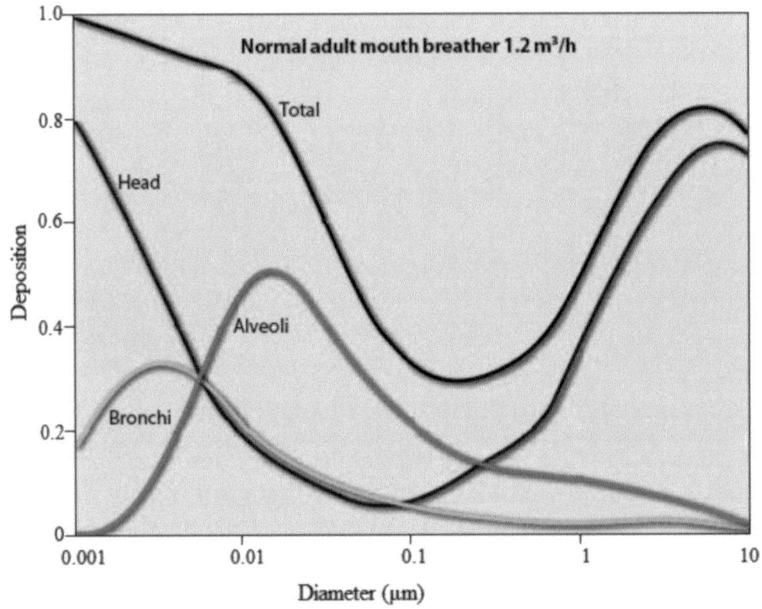

Fig. 5. Respiratory tract deposition probability of inhaled particles

The few epidemiological studies that have addressed this important question specifically suggest that combustion sources are particularly important for health. Toxicological studies have also pointed to primary combustion-derived particles as having a higher toxic potential. These particles are often rich in transition metals and organic compounds, and also have a relatively high surface area. By contrast, several other single components of the PM mixture (e.g. ammonium salts, chlorides, sulfates, nitrates and wind-blown dust such as silicate clays) have been shown to have a lower toxicity in laboratory studies (Schwartz, J. et al. 1996).

Despite these differences found among constituents studied under laboratory conditions, it is currently not possible to quantify the contributions from different sources and different PM components to the effects on health caused by exposure to ambient PM. Nevertheless, it seems reasonable to include in abatement efforts those sources/constituents that have been shown to be critical, such as emissions from diesel engines.

Many studies have found that fine particles (usually measured as $PM_{2.5}$) have serious effects on health, such as increases in mortality rates and in emergency hospital admissions for cardiovascular and respiratory reasons. Thus there is good reason to reduce exposure to such particles. Coarse particles (usually defined as the difference between PM_{10} and $PM_{2.5}$) seem to have effects on, for example, hospital admissions for respiratory illness, but their effect on mortality is less clear. A few studies suggest that fine PM is more biologically active than coarse PM (defined as particles between 2.5 and 10 µm in size) (Klemm, et al 2000; Schwartz, J. & Neas L. M. 2000; R.W. Atkinson et al., 2000; F. Dominici et al 2007).

Nevertheless, there is sufficient concern to consider reducing exposure to coarse particles as well as to fine particles. Up to now, coarse and fine particles have been evaluated and regulated together, as the focus has been on PM_{10}. However, the two types have different sources and may have different effects, and tend to be poorly correlated in the air. The systematic review therefore recommended that consideration be given to assessing and controlling coarse as well as fine PM. Similarly, ultrafine particles are different in composition, and probably to some extent in effect, from fine and coarse particles.

Annual mean level	PM_{10} (μg/m³)	$PM_{2.5}$ (μg/m³)	Basis for the selected level
WHO interim target-1 (IT-1)	70	35	These levels are estimated to be associated with about 15% higher long-term mortality than at AQG
WHO interim target-2 (IT-2)	50	25	In addition on the other health benefits, these levels lower risk of premature mortality by approximately 6% [2-11%] compared to WHO-IT1
WHO interim target-3 (IT-3)	30	15	In addition on the other health benefits, these levels lower risk of premature mortality by approximately 6% [2-11%] compared to WHO-IT2 levels
WHO Air quality guidelines (QAG)	20	10	These are the lowest levels at which total, cardiopulmonary and lung cancer mortality have been shown to increase with more than 95% confidence in response to $PM_{2.5}$ in the ACS study. The use of $PM_{2.5}$ guideline is preferred.

Table 9. Air Quality guidelines for PM (annual)

Nevertheless, their effect on human health has been insufficiently studied to permit a quantitative evaluation of the risks to health of exposure to such particles. Multi-city studies of 29 cities in Europe and 20 cities in the United States (Health Effects Institute, 2004) reported short-term mortality effects for PM10 of 0.62% and 0.46% for every 10 μg/m³ respectively. A meta-analysis of 29 cities from outside Western Europe and North America reported an effect of 0.5%. A meta-analysis confined to Asian cities reported an effect of 0.49%. This suggests that the health risks for PM10 are likely to be similar in cities in developed and underdeveloped countries at around 0.5%. Therefore, a concentration of 150 μg/m³ would relate to roughly a 5% increase in daily mortality, an impact that would be of significant concern. Tables 9 and 10 illustrate the WHO guidelines for two different averaging times.

24-hour mean level*	PM_{10} (μg/m^3)	$PM_{2.5}$ (μg/m^3)	Basis for the selected level
WHO interim target-1 (IT-1)	150	75	Based on published risk
WHO interim target-2 (IT-2)*	100	50	Based on published risk coefficients from multicentre studies and meta-analyses (about 2.5% increase in short-term mortality over AQG
WHO interim target-3 (IT-3)**	75	37.5	About 1.2% increase in short-term mortality over AQG
WHO Air quality guidelines (AQG)	50	25	Based on relation between 24-hour and annual PM level

Table 10. Air Quality guidelines for PM (24-hr), *99th percentile (3 days/year), ** for management purpose, based on annual average guideline values; precise number to be determined on basis of local frequency distribution of daily means

4.5 Health effects due to nitrous oxide (NO_x)

NO_2 acts mainly as an irritant affecting the mucosa of the eyes, nose, throat, and respiratory tract. Extremely high-dose exposure (as in a building fire) to NO_2 may result in pulmonary edema and diffuse lung injury. Continued exposure to high NO_2 levels can contribute to the development of acute or chronic bronchitis. Low level NO_2 exposure may cause increased bronchial reactivity in some asthmatics, decreased lung function in patients with chronic obstructive pulmonary disease and increased risk of respiratory infections, especially in young children.

Short-Term	*Long-Term*
Effects on pulmonary function, particularly in asthmatics	Reduction in lung function
Increase in airway allergic inflammatory reactions	Increased probability of respiratory symptoms
Increase in hospital admissions	
Increase in mortality	

Table 11. Health Effects due to NO_x (Finlayson – Pitts & Pitts 1999)

Guidelines are established as follows (WHO, 2005):

NO_2 concentration: 40 μg/m^3 for annual mean,

and NO_2 concentration: 200 μg/m^3 for 1-hour mean

Effects of NO_2 are more difficult to isolate independently because NO_2 is an important constituent of combustion-generated air pollution and is highly corelated with other primary and secondary combustion products. No mortality or illness statistics can be associated yet based on lack of evidence.

4.6 Health effects due to ozone (O_3)

Recent epidemiological studies have strengthened the evidence that there are short-term O_3 effects on mortality and respiratory morbidity and provided further information on exposure response relationships and effect modification. Based on a meta-analysis of studies published during the period between 1996 and 2001 on short-term effects of O_3 on all non-accidental causes of death in all ages (or older than 65 years), significant increase of the risk of dying (between 0.2% and 0.6% per each increase in 10 µg/m³) was shown (Royal 2007).

The National Morbidity Mortality Air Pollution Study (NMMAPS) study, reported a significant effect of O_3 during the summer season, of 0.41 % increase in mortality associated with an increase of 10 ppb (20 µg/m³) in daily O_3 "same-day" concentrations. Ozone daily levels were associated with hospital respiratory admissions at all ages in most of the studies using 8-hour measures and also in many of the studies using other averaging periods.

The magnitude of the association was slightly larger than that obtained for mortality (0.5 to 0.7% increases in admissions per increase of 10 µg/m³ in O_3. Studies on admissions for asthma in children did not find conclusive associations with any O_3 measurement (Paul Hawken, et al 1999, Reeves H & Lenoir F.2005). The effects of long-term exposure to Ozone are much less known. Table 12 provides a summary of health effects related to Ozone. Table 13 presents the WHO guidelines for 8-hr averaging of Ozone concentrations.

Short-Term	*Long-Term*
Adverse effects on pulmonary function	Reduction in lung function development
Lung inflammatory reactions	
Adverse effects on respiratory symptoms	
Increase in medication usage	
Increase in hospital admissions	
Increase in mortality	

Table 12. Health Effects due to O_3

	Daily maximum 8-hour mean (µg/m³)	**Effects at the selected ozone level**
High level	240	Significant health effects, substantial proportion of vulnerable population affected
WHO interim target-1 (IT-1)	160	Important health effects, an intermediate target for populations with ozone concentrations above this level. Does not provide adequate protection of public health. Rationale: Lower level of 6.6-hour chamber exposures of healthy exercising young adults, which show physiological and inflammatory lung effects. Ambient level at various summer camp studies showing effects on health of children Estimated 3-5% increase in daily mortality* (based on findings of daily time-series studies)

WHO Air quality guidelines (AQG)	100	This concentration will provide adequate protection of public health, though some health effects may occur below this level. Rationale: Estimated 1-2% increase in daily mortality* (based on findings of daily time-series studies) Extrapolation from chamber and field studies based on the likelihood that real-life exposure tends to be repetitive and chamber studies do not study highly sensitive or clinically compromised subjects, or children Likelihood that ambient ozone is a marker for related oxidants

Table 13. Ozone Air Quality guidelines, *Deaths attributable to ozone concentrations above estimated baseline of 70 μg/m^3. Based on range of 0.3 to 0.5% increase in daily mortality for 10 μg/m^3.

5. Indoor air quality

The quality of air inside homes, buildings, schools, day care centers, etc is very important. The quality of the air that we breathe can have important effects on our health and quality of life. However, we breathe all time when we are outdoors or indoors. We are used to thinking of the outdoor environment to be safe from air pollution. It is known that during smog or dusty air people are advised to stay indoors. Yet new research, in particular research for the astronauts from the National Aeronautics and Space Administration (NASA), faced the problem of indoor air pollution and began extensive studies on treating and recycling air in the chambers. These studies lead to the problem of indoor air pollution. They discovered that the indoor environment may be as much as ten times more polluted than the outdoor environment. However, as early as 1950 Dr. T.G. Randolph (Wolverton, 1996) became one of the first medical doctors to link indoor air pollution with allergies and other chronic diseases. Still today millions of people fail to realize the serious nature of the problem. Today people living in cities and in industrialized environments spend as much as 80% of their lives indoors fail to recognize this problem. Exposure to indoor air pollutants, which are many as we will see later, correlates to an increase in the number of allergic reactions, as well as to chronic diseases due to toxic substances. NASA scientists started to study the development of sustainable indoor ecological life- support facilities. The NASA scientists soon discovered that houseplants could purify air in sealed test-chambers. As many people become concerned about the direct association of indoor environment and their health, the green revolution will grow. If we stress the importance of indoor air quality and to relate our existence to a symbiotic and beneficial relationship with the animals and plants of our nature then we will be closer to our living world.

5.1 House plants and indoor air quality

Evidence is given to show how houseplants can become a necessary component of healthy buildings whether houses or offices and how houseplants can improve the indoor air quality. Houseplants are capable of removing toxic chemical vapors. Low relative humidity levels, below 35 percent are also associated with poor IAQ. Frequent colds and allergic asthma during the cold winter months are often caused by low relative humidity. Emissions from modern materials used to construct home or office furniture from pressed wood products or fiberboard, which often replace natural wood in building construction, as well as wall- to- wall carpeting are synthetic materials and are held together with glues and resins. Furthermore, a number of electronic devices are found in our homes today, such as radios, televisions, etc. for our pleasure are known to emit various organic compounds. The synthetic materials release hundreds of volatile organic materials (VOCs) into the indoor air. Compounds that may be found in the air of indoor houses, buildings and offices may be formaldehyde, xylene, toluene, benzene, chloroform, alcohols, acetone, etc. Humans are also a source of indoor air pollutants especially in closed and poorly ventilated areas. In addition to carbon dioxide humans release many volatile substances, in the atmosphere, which are called "biofluents", such as, ethyl alcohol, methyl alcohol, acetone, ammonia, etc.

Thus, sealed buildings and synthetic furnishings are the main sources of indoor air pollution a phenomenon known as "sick building syndrome"(SBS) with some common symptoms, i.e. allergies, asthma, eye, nose and throat irritation, fatigue, headache, respiratory congestion, sinus congestion and others. Some also include lung cancer from asbestos exposure.

5.2 Epidimiologic studies

The epidemiological study into symptoms among office workers has produced many important results, which are conflicting due to methodologic issues in the interpretation of the epidemiological findings (Mendell, 1993). The environmental factors that were found increased symptoms with air conditioning, carpets, video display terminals, etc. The ventilation rates near or below 10 liters/second/person decreased symptoms. Personal factors, such as female gender, job stress/dissatisfaction and allergies/asthma were also studied and showed increased symptoms with the above factors. The evidence suggested that work related symptoms among office workers were relatively common. Indoor exposure and problems due to this exposure could be reduced if prevention of building related symptoms may be eliminated with appropriate design, operation and maintenance practices, such as ventilation rates (Zuraimi, 2010).

In another study microbial indoor air quality and respiratory symptoms of children in schools with visible moisture and mold problems showed that school buildings of concrete/brick developed fungi concentration, but not in wooden school buildings (Meklin et al, 2002). There are more epidemiological studies, which indicate that there are risks associated with elevated air fine particle concentrations (Mullen et al. 2011, Pope & Dockey, 2006).

Potential health risks may result from environmental exposure to ultrafine particles (< 0.1 μm diameter) in particular exposure in school classrooms. It was found that average indoor levels were higher when classrooms were occupied than when they were unoccupied due to ultrafine particle concentrations (Mullen et al. 2011).

A multi location indoor study in air settled dust showed abundance of orthophosphate and phthalate esters (Bergh et al. 2011). Both groups of chemicals are semi volatile compounds and they are additives in plastic materials, which are used into indoor environment as industrial chemicals emanating from furniture in general. These chemicals were found in private homes, day care centers, and workplaces in the Stockholm area. The phthalate esters were 10 times higher than the orthophosphate esters. Especially high levels of tributoxyethyl phosphate were found in the day care centers and high levels of diethylhexyl phthalate in dust.

6. References

Atkinson, R.W. et al., (2000) Acute Effects of Particulate Air Pollution on Respiratory Admissions, Am. J. Respir. Crit. Care Med. 164/10, 1860-1866,

Arribas-Monzón, F.; Rabanaque, M.J.; Martos, C.; Abad, J.M.; Alcatá-Nalvaiz, T. & Navarro-Elipe, M. (2001) Effects of air pollution on daily mortality in Zaragoza, Spain, 1991-1995, Salud Pública México. Pp. *43* 1-8.

Basu R. & Samet J. M. (2002) Relation between Elevated Ambient Temperature and Mortality: A Review of the Epidemiologic Evidenc, *Epidemiol. Rev.*, 24(2), 190-202

Bergh, C.; Torgrip, R.; Emenius, G. & Östman, C. (2011). Organophosphate and phthalate esters in air and settled dust- a multi-location indoor study. Indoor Air; 21; 67-76.

Coull, B.A.; Schwartz, J. & Wand, MP. (2001). respiratory health and air pollution: additive mixed model analysis, *Biostatistics*, 2/3; 337-349

Council on Environmental Quality και *Airport Environmental Handbook,* Order 5050.4A, National Environmental Policy Act (NEPA). Airports Orders.

Dominici, F. (2002) Invited Commentary: Air Pollution and Health –What Can We Leaarn from a Hierarchical Approach, *Am. J. Epidemiol.*, 155, , 1-15

Dominici, F. et al, (2007). Particulate Air Pollution and Mortality in the United States : Did the Risks Change fron 1987 to 2000, Am. J. Epidem., 166 (8), 880-888,

Filliger P., Puybonnieux-Texier V. and Schneider J. (1999). Health costs due to road traffic-related air pollution. An impact assessment project of Austria, France and Switzerland. PM10 population exposure. Technical report on air pollution. GVF-Report nr 326 - TEH05.

Filleul L, Baldi I., Dartigues J.-F., and Tessier J.-F., Risk factors among elderly for short term deaths related to high levels of air pollution, *Occup. Environ. Med.*, 60(9) 2003, 684-688

Finlayson - Pitts B. J. & Pitts Jr., J. N., (1999). Chemistry of the Upper and Lower Atmosphere : Theory, Experiments and Applications, Hardcover, Academic Press

Hawken, P.; Lovins, A.B. & Hunter, L. (1999). Natural Capitalism: Creating the Next Industrial Revolution, Little Brown Company.

Health Effects Institute, 2004, International Oversight Committee, Special reports, Particulate Matter

Health Canada (2005). Estimated Number of Excess Deaths in Canada due To Air Pollution", S. Judek, B. jessiman, D. Stieb, Air Health Effects Division, Health Canada and R. Vet, Meteorological Service of Canada, Environment Canada, April.

Jerrett M., Burnett R. T., Brook J., Kanaroglou P., Giovis C., Finkelstein N., & Hutchison B. 200. Do socioeconomic characteristics modify the short term association between air pollution and mortality? Evidence from a zonal time series in Hamilton, Canada *J. Epidemiol, Community Health*, 58(1), 31-40

Judek, S.; jessiman, B. ; Stieb, D. & Vet, Health R. (2005). Estimated Number of Excess Deaths in Canada due To Air Pollution, Air Health Effects Division, Health Canada Meteorological Service of Canada, *Environment Canada*, April.

Katsouyanni, K. (2003). Ambient air pollution and health, British Medical Bulletin, *68* 143-156.

Klemm, R. J. ET AL. (2000).Is daily mortality associated specifically with fine particles? Data reconstruction and replication of analyses. Journal of the air and waste management association, 50: 1215–1222

Kotzias, D. Human exposure research, Needs and Approaches, 8th FECS Conference. 2003, 13.

Kunzli N., Medina S., Kaiser R., Quenel P., Horak F. Jr. & Studnicka M. (2001). Assessment of Deaths Attributable to Air Pollution: Should We Use Risk Estimates based on Time Series or on Cohort Studies? *Am. J. Epidemiol.*, 153, 1050 - 1055

Le Tertre A., Medina S., Samolli E., Forsberg B., Michelozzi P., Boumghar A., Vonk J. M., Bellini A., Atkinson R., Ayres J. G., Sunyer J., Schwartz J., and Katsouyanni K., Short-term effects of particulate air pollution on cardiovascular diseases in eight European cities, *J. Epidem. Commun. Health*, 56 , 2002, 773-779

Mcdonnell, w.f. et al. (2000). Relationships of mortality with the fine and coarse fractions of long-term ambient PM10 concentrations in nonsmokers. J Exposure Analysis Environmental Epidemiology, 10: 427–436.

Meklin, T.; Husman, T.; Vepsäläinen, T.; Vahteristo, M. Koivisto, J.; Halla-Aho, J.; Hyvärinen, A.; Moschandreas, D. & Nevalainen, A. (2002). Indoor Air. 12; 175-183.

Mendell, M. (1993). Non-specific symptoms in office workers: A review and summary of epidemiologic literature, Indoor Air. 3; 227-236.

Mullen, N.A.; Bhangar, S.; Hering, S.V.; Kreisberg, N.M. & Nazarof. W.W. (2011). Indoor Air. Ultrafine particle concentrations and exposures in six elementary school classrooms in northern California. 21; 77-8

Rita K. Seethaler, K. Künzli N., Sommer H., Chanel O., Herry M5, Masson S., Vernaud J-C., Filliger P., Horak F.Jr., Kaiser R., Medina S., Puybonnieux-Texier V., Quénel P., Schneider J., Studnicka M., Heldstab J. (2003). Economic costs of air pollution-related health impacts, An Impact Assessment Project of Austria, France and Switzerland, *Clean Air & Environmental Quality*, Vol. 37, No. 1, 35-43

Reeves H, Lenoir F., Mal de Terre, Editions du Seuil, 2005.

Royal Commission on Environmental Pollution (2007). The Environmental Effects of Civil Aircraft in Flight. 22 March, Report.

Samet, J.M.; Dominici, F.; Curriero, F.C.; Coursac, I. & Zeger S. L. (2000). Fine particulate air pollution and mortality in 20 U.S. cities 1987–1994. New Eng J Me., 343: 1742–1749

Samoli E., Touloumi G., Zanobetti A., Le Tertre A., Schindler C., Atkinson R., Vonk J., Rossi G., Saez M., Rabczenko D., Schwartz J. & Katsouyanni K.(2003) Investigating the dose-response relation between air pollution and total total mortality in the APHEA-2 multicity project, *Occup Environ Med. 60:977-982*

Samet J.M. , Zeger, S.L., Dominici F., Curriero F., Coursac I., Dockery D.W. & Schwartz J., (2000). The National Morbidity, Mortality, and Air Pollution Study. Part II: Morbidity and mortality from air pollution in the United States. Health Effects Institute. North Andover MA: Flagship Press. 94 Part II:1–82.

Schwartz, J. Dockery, D.W., Neas, L.M. (1996). Is daily mortality associated specifically with fine particles? Journal of the air and waste management association, 46: 927–939

Schwartz, J. & Neas L. M. (2000) Fine particles are more strongly associated than coarse particles with acute respiratory health effects in schoolchildren. Epidemiology, 11: 6–10

Sommer, H., Seethaler, R., Chanel, O., Herry, M., Masson, S., Vergnaud, J.-C. (1999). Health costs due to road traffic-related air pollution. An impact assessment project of Austria, France and Switzerland. Economic evaluation. Technical report on economy. GVF-Report nr 326 - TEH07 Federal Department for Environment, Transport, Energy and Communications Bureau for Transport Studies, Bern.

Prepared for the Third World Health Organisation Ministerial Conference of Environment and Health, London, 1999.

Sunyer J. and Basagana X., Particles, and not gases, are associated with the risk of death in patients with chronic obstructive pulmonary disease, *Int. J. Epidemiol.*, 30, 2001; 1138-1140

Theophanides, M.; Anastassopoulou, J. & Theophanides, T. (2002). A statistical study of disease-related mortalities due to environmental pollutants in Kavala, Greece, In Environmental science and pollution research, 8th FECS Conference on chemistry and the environment, 44.

Theophanides, T.; Vassilakos, Ch.; Anastassopoulou, J.; Maggos, T.; Hatzianestis, J.. & Bartzis, I. (2002). Chemical Characterization of VOCs in Nea Karvali area, Kavala, Greece, In *Envirommental science and pollution research, 8th FECS Conference on chemistry and the environment*, 45-46.

Theophanides, M.; Anastassopoulou, J.; Vasilakos, Ch.; Maggos, Th. & Theophanides, T. (2007). Mortality and Pollution in Several Greek Cities, J. Envir. Scien Health, Part A, 42/6; 741-746.

Theophanides, M. & Theophanides, T. Human environmental engineering and sustainability, in M. Theophanides and T. Theophanides Eds., Environmental Engineering and Economics, ATINER, Athens, Greece, pp. 1-8.

Theophanides, M. & Anastassopoulou, J. (2009). Simulation of Air Pollution and Health Effects, In: in *Environmental Awarness and Management*, eds, T. Theophanides and M. Theophanides, ATINER, Athens, Greece, pp. 29-37

Touloumi, G.; Pocock, SJ.; Katsouyianni, K. & Trichopoulos, D. Short-term effects of air pollutions on daily mortality in Athens: a time- series analysis, Int. J. Epidemiol. 1994, *23* 957-967.

The MACBETH project (1999): Monitoring of atmospheric concentrations of benzene in European towns and homes, *EU LIFE project 96ENV/IT070*

Tunnicliffe W. S., Harrison R.M., Kelly F. J., Dunster C., and Ayres J. G., The effect of sulfurous air pollutant exposures on symptoms, lung function, exhaled nitric oxide, and nasal epithelial lining fluid antioxidant concentrations in normal and asthmatic adults, *Occup. Environ. Med.* 60(11), 2003, 15-15

Tobias, A.; Saez, M.; Daniels, M. J,; Dominici, F.; Zeger S. L. & Samet J. M., RE (2001). Estimating Particulate Matter - Mortality Dose - Response Curves And Threshold Levels: An Analysis Of Daily Time - Series For The 20 Largest Us Cities, *Am. J. Epidemiol.*, 153, 1027 - 1028

WHO (2002). Systematic review of health aspects of AQ Europe. An overview of the St. George's project. A systematic review of the epidemiological literature on the short-term health effects of outdoor air pollution, St. George's Hospital, London, United Kingdom.

Wolverton, B.C. (1996).How to Grow Fresh Air: Houseplants that purify your home or office, Penguin Books.

Yang, C-Y.; Chang, C-C.; Chuang, H-Y.; Tsai, S-S.; Wu, T-N.; Ho, C-K. Relationship between air pollution and daily mortality in a subtropical city: Taipei, Taiwan, Environm. Int. 2004, *30* 519-523.

Zuraimi, M. (20101). Is ventilation dust useful? A review of scientific evidence. Indoor Air. 20; 445-457.

2

Development and Evaluation of a Dispersion Model to Predict Downwind Concentrations of Particulate Emissions from Land Application of Class B Biosolids in Unstable Conditions

Abhishek Bhat, Ashok Kumar and Kevin Czajkowski[1]
Department of Civil Engineering, The University of Toledo, Toledo,
[1]Department of Geography and Planning,
The University of Toledo, Toledo,
USA

1. Introduction

The term, biosolids, is generally used to refer to those waste products that have been stabilized by treatment of the sewage sludge for beneficial reuse through appropriate management (Davis, 2002). The agronomic and environmental benefits from the organic material and fertilizing elements contained in the biosolids are essential for maintaining soil fertility. This has been a major reason for the application of biosolids on the agricultural fields. These biosolids reused for land application on agricultural fields has potential benefits. Davis (2002) in his study described the following benefits:

1. The land application of biosolids is mainly used to improve the soil quality. The organic matter from the soil can be built. Water retention, soil stabilization, and reduced soil erosion are some of the other benefits.
2. Applied biosolids can partially or completely substitute commercial fertilizer. These biosolids contain nutrients present in conventional fertilizer including nitrogen, phosphate, and other additive elements.
3. The application of biosolids or reuse of biosolids reduces the quantity of waste required to be disposed in landfills. This reduces the pollution due to landfills, leachates, etc.

The process of land application of biosolids on agricultural land has been carried out for generations. The agricultural activities related to the land application of biosolids aerosolize particulate matter. The United States Environmental Protection Agency (US EPA) regulates particulate matter as a "criteria pollutant". The particulate matter emitted during various agricultural activities impact air quality. The particulate matter generated from agricultural activities includes dust from the fields and dust generated from agricultural activities. The particulate matter emitted from the agricultural activities can contain bioaerosols, endotoxins, and pathogens. The airborne particles consisting of or originating from the microorganism are called bioaerosols. Bioaerosols containing pathogenic bacteria and harmful microorganisms accompanied with handling and the application process could harm the public health and environment. Modeling transport and dispersion of the

particulate matter emitted during the land application of biosolids is important to predict the downwind concentrations and in turn to predict the risk.
The objective of this chapter is to model the particulate matter released during and after the application of biosolids based on the data collected during the field study. The efforts include a derivation of solution to the convective-diffusion equation incorporating wind shear.

2. Literature review

Emissions of particulate matter during the application of these biosolids were studied by various researchers. Paez-Rubio et al. (2006) studied the composition of these particulate matters and determined the emission rates due to disking activity. The researchers used arrayed samplers to estimate the vertical source aerosol concentration, which were used to calculate the plume. The different constituents of the biosolids and their emission rates were reported in the study.
Brooks et al. (2005) derived an empirical equation to estimate the bioaerosols risk infection to residents adjacent to the land that is applied with biosolids. For this study, a coliphage MS-2 and Escherichia coli organisms were aerosolized after adding them to water within a biosolids spray application truck. Then the downwind concentration of these microorganisms was measured at various distances ranging from 2 m to 70 m. The data were taken downwind of the sprayer and were used to derive an empirical equation. The limitation of this study is that the authors used a simplistic regression model to determine the transport. US EPA's SCREEN 3 dispersion model was used to predict the downwind concentrations of particulate aerosols in the study by Taha et al. (2005). The emission rates in this study were determined by the wind tunnel experiments conducted on the surface of the static compost windrows. In a similar study, Dowd et al. (2000) predicted the downwind concentration of airborne viruses from a biosolids placement site. The study incorporated a modified Gaussian equation to quantify the downwind concentrations in an area undergoing the land application of biosolids. The model was used to predict the downwind concentration of microorganisms from an area source by taking into account the length and the width of the agricultural field.
A major difference between a conventional source of particulate matter and an agricultural source is that the later is a ground level source. Conventionally the wind velocity used in the downwind concentration calculated by researchers was used as an average velocity which was assumed to be constant over the vertical stretch of the plume. In real conditions, near the ground level, the magnitude of velocity changes with the change in vertical height. A vertical shear layer is formed and the velocity varies at a rapid rate near the ground. Thus the concentrations predicted can show large variations if the wind shear is not taken into account during dispersion. Kumar and Bhat (2008) discuss a possible generic model for transport and dispersion of particulate matters incorporating wind shear (magnitude shear only) near the ground. There is a need to understand and apply the knowledge of dispersion modeling to particulate fate and transport. It is important to develop a general screening model to predict downwind concentrations. The account for wind shear near the ground needs to be studied and incorporated in the existing models. The book entitled "Micrometeorology" by Sutton (1953) gives a solution using the variable eddy diffusivity and wind speed for steady state two-dimensional convective-diffusion equation representing the diffusion from an infinite line source. Kumar and Bhat (2008) extended the

analytical solution given by Sutton (1953) to predict the concentrations for ground level area sources. The new model has been evaluated using the data collected in 2009 and the regression equation given by Brooks et al. (2005) based on their field work.

3. Field sampling study

In the summer of 2009, a field study was conducted to collect particles emitted during the land application of biosolids. Particle emissions were collected for three days during the application (application), and for two days after the application (post-application) of biosolids. An agricultural field, scheduled for application of Class B biosolids in Northwest Ohio was selected for the sampling. The biosolids were applied on this field by injection method.
Particle samples were collected via the use of two GRIMM 1.108 aerosol samplers operating at airflow of 1.2 l/minute. The gravimetric data in 16 channels over the size range 0.23 μm < d < 20 μm was collected for a total of six hours every sampling day. The samplers were placed onto specially arranged tables raised to a height so that the intake nozzle was at average human breathing height of 1.5 m. Two sampling stations, one station inside the field and one outside were selected. The location of the outside sampling station at 10 m downwind from edge of the field was changed to 20 m downwind after first three hours of sampling keeping the location of the inside station same throughout the sampling. The monitors were reoriented in the direction of the wind, if needed. The weather data were collected using a portable weather station at both sampling locations inside and outside. The atmospheric parameters defining the atmospheric stability for each hour of sampling on each sampling day are presented in Table 1. The location of outside concentration monitoring station for each hour is also noted. The atmospheric stability for almost all sampling days was slightly unstable to moderately unstable. On one occasion it was slightly unstable to neutral.

Date	Time	Concentration Monitor Location from Edge	Wind Velocity (m/s)	Wind Condition	Cloud Cover (in tenth)	Daily Solar Radiation (W/m^2)	Atmospheric Stability using P-G Method*
Application August 21, 2009	09:25–10:25	@ 10 m	5.81	Very High	0	755	C
	10:25-11:25		8.56				C
	11:25-12:25		8.59				C
	12:25-13:25	@ 20 m	8.93				C
	13:25-14:25		8.85				C
	14:25-15:25		8.64				C
Application August 24, 2009	09:17–10:17	@ 10 m	0.27	Calm	4	373	B
	10:17-11:17		0.33				B
	11:17-12:17		0.25				B
	12:17-13:17	@ 20 m	0.68				B

	13:17-14:17		0.60				B
	14:17-15:17		0.41				B
Application August 26, 2009	08:00-09:00	@ 10 m	3.46	Low	8	288	C
	9:00-10:00		3.73				C
	10:00-11:00		2.94				C
	11:00-12:00	@ 20 m	2.27				C
	12:00-13:00		1.91				B
	13:00-14:00		2.39				C
Post-Application Sept. 24, 2009	08:40-09:40	@ 10 m	0.14	Calm	8	327	B
	09:40-10:40		0.14				B
	10:40-11:40		0.25				B
	11:40-12:40	@ 20 m	0.40				B
	12:40-13:40		0.32				B
	13:40-14:40		0.13				B
Post-Application Sept. 25, 2009	08:30-09:40	@ 10 m	4.07	High	5	541	C-D
	09:30-10:30		5.26				C-D
	10:30-11:30		5.87				C-D
	11:30-12:30	@ 20 m	5.45				C-D
	12:30-13:30		6.13				D
	13:30-14:30		5.78				C-D

*B: Moderately Unstable; C: Slightly Unstable; D: Neutral

Table 1. Atmospheric Conditions Observed on Each Sampling Day

The concentration data collected during the application and the post-application was processed using Microsoft Office 2010 Excel sheets. Hourly average concentrations for each day were calculated. Based on the average wind velocities (u) measured, sampling days were divided into three windy conditions; low wind condition (0.5 m/s < u < 3 m/s), high wind condition (3 m/s < u < 6 m/s), and very high wind condition (u > 6 m/s) (see Table 1). The data collected at the inside station represented the emissions generated during the agricultural activities. The vertical profiles of particle dispersion inside the agricultural field during and after sludge application analyzed by Akbar et al. (2011) were used to develop a set of emission rate equations. Hourly emission rates (Q) for each sampling day were calculated using these emission rate equations. The data collected at the outside sampling stations was used as the downwind concentration (C).

4. Model development

4.1 Shear layer model development

There are different equations available in literature for the dispersion of a ground level release of a pollutant. However, none of the reported equations tackles the problem of wind shear near the ground. This part focuses on deriving the analytical solution from the convection-diffusion equation using vertical velocity profile. The following assumptions are used in deriving the equation:

1. The wind direction is always perpendicular to the field.
2. The dispersion is of the non-fumigation type.

The velocity profile with height above the ground level is assumed to be the same for all downwind distances. The magnitude of the wind velocity near the ground level changes rapidly. Therefore, for the ground level discharge of the pollutant, it is very important that the variation of the wind velocity magnitude is incorporated in the dispersion and transport equation.

The model uses the equation for C(x,z) given by Sutton (1953):

$$C(x,z) = \frac{Q}{u_1 * \Gamma(S)} * \left[\frac{u_1}{(m-n+2)^2 * K_1 * x}\right] * \exp[-u_1 * \frac{z^{m-n+2}}{((m-n+2)^2 * K_1 * x)}] \quad (1)$$

where,

C(x,z): Downwind concentration (unit/m^3)

x: Downwind distance (m)

z: Vertical distance (m)

Q: Emission rate of pollutants (unit/sec)

u_1: Wind velocity reference height Z_1 by the power law $u(z) = u_1 * \left(\frac{z}{z_1}\right)^m$ (2)

K_1: Diffusivity constant reference height Z_1 given by $K(z) = K_1 * \left(\frac{z}{z_1}\right)^n$ (3)

n: Exponent of power law velocity profile

m: Exponent of eddy diffusivity profile where, m = 1 – n

s: Stability parameter based on m and n ($s = \frac{m+1}{m-n+2}$)

Γ(s): Gamma function of s

The Equation (1) is integrated from x-(X/2) to x+(X/2) for a strip source with width X, and infinite length having the origin of x ordinate at the center of the strip to obtain the concentration from the strip. The integration gives following formulae given by Kumar and Bhat (2008).

$$C(x,z) = Q * \frac{z^{a^{S-1}}}{A} * \left[\frac{A + x^{1-s} * B * \exp\left(\frac{B}{x}\right) + D}{S-1}\right]_{x-\left(\frac{X}{2}\right)}^{x+\left(\frac{X}{2}\right)} \quad (4)$$

where,

$$A = \gamma(s) \quad (5)$$

$$B = -u_1 * \frac{z^a}{a^2 * K_1} \quad (6)$$

$$D = \gamma(s, \left(-\frac{B}{x}\right)) \tag{7}$$

$$a = (m - n + 2) \tag{8}$$

The total concentration of the pollutant is given by following equation after considering i number of strips in the area source.

$$C(x,z) = \sum_1^i Q * \frac{z^{a^{S-1}}}{A} * \left[\frac{A + x^{1-S} * B * \exp\left(\frac{B}{x_i}\right) + D}{S-1}\right]_{x_i - \left(\frac{X}{2}\right)}^{x_i + \left(\frac{X}{2}\right)} \quad \text{for } z > 0 \quad \text{(9-a)}$$

$$C(x) = \sum_1^i \left[\frac{Q}{\Gamma(S) * (m+n-2)^2 * K_1} * \ln(x_i)\right]_{x_i - \frac{X}{2}}^{x_i + \frac{X}{2}} \quad \text{for } z = 0 \quad \text{(9-b)}$$

The value of x_i is calculated using

$$x_i = x_d + \frac{X}{2} \text{ (for i = 1)} \tag{10}$$

and

$$x_i = x_{i-1} + X \text{ (for i > 1)} \tag{11}$$

where, x_d is the downwind distance of monitoring station from the edge of the field.
The Equation (9-a) computes the concentration of the pollutant at chosen breathing level while the downwind concentration at the ground is computed using Equation (9-b). These Equations (9-a) and (9-b) were modeled into an Excel spreadsheet as the Shear Layer Model as part of Bioaerosols Dispersion and Risk Model spreadsheet (BDRM 1.01). The programming is done in a way so that the calculated concentrations are from the edge of the field for different downwind distances. The development of BDRM spreadsheet is discussed in Kumar and Bhat (2008).

5. Model evaluation

The evaluation of shear layer model involved two major steps: 1. the predicted concentrations from the shear layer model were compared to the measured concentration data from field study and 2. the model was evaluated using the limited data available in the literature. In each step, the predicted data were evaluated using the calculated statistical parameters.

5.1 Model evaluation using measured data

Multiple runs of the shear layer model were carried out to simulate characteristics of each sampling day. Since the shear layer model was not developed for the calm conditions, only sampling days with different windy conditions were modeled. The turbulence parameters used to simulate the atmospheric turbulence in the shear layer model are presented in Table 2. The values of n were based on urban and rural exponents used in the air quality models developed by the US EPA and K_1 was calculated using the equations compiled by Kumar (1977). The predicted concentrations and the measured concentrations were formatted into a Microsoft Excel spreadsheet to obtain average hourly concentrations. The predicted concentrations from the shear layer model were compared with the measured concentrations (Figure 1). Visible comparison were enabled by plotting the measured vs. predicted data on the same plot. It was found that the shear layer model over predicts the concentration for all windy conditions except for few data points.

Model Input	Neutral	Unstable	Stable
m	0.85	0.8	0.7
n	0.15	0.2	0.3
K_1 (m^2/sec)	8	28.43	0.993

Table 2. Input used for the Shear Layer Model

The statistical evaluation based on the work of Hanna et al. (1993), Gudivaka and Kumar (1990), Riswadkar and Kumar (1994) and Kumar et al. (2006), was used in this study. In order to determine the significance of the evaluation of the model, four statistical parameters; normalized mean square error (NMSE), fractional bias (FB), correlation coefficient (R), and geometric mean bias (MG) were calculated.

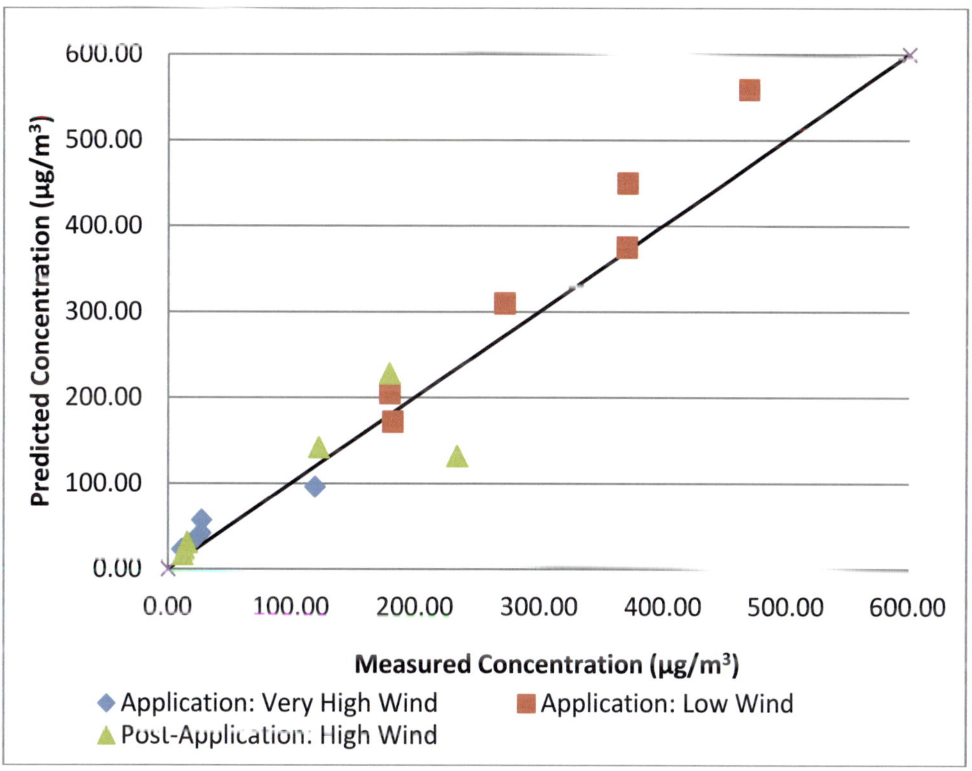

Fig. 1. Measured vs. Predicted Concentration

The normalized mean square error (NMSE) is given by the formula,

$$NMSE = \frac{\overline{(C_O - C_P)^2}}{\overline{C_0} \times \overline{C_P}} \tag{12}$$

The fractional bias (FB) is given by the formula,

$$FB = 2 \times \left(\frac{\overline{C_0} - \overline{C_P}}{\overline{C_0} + \overline{C_P}} \right) \tag{13}$$

The correlation coefficient (R) is given by the formula,

$$r = \frac{\overline{\left(C_0 - \overline{C_0}\right)\left(C_P - \overline{C_P}\right)}}{\sigma_{C_P} \sigma_{C_0}} \tag{14}$$

And the geometric mean (MG) bias is calculated by the formula,

$$MG = \exp\left(\overline{\ln C_0} - \overline{\ln C_P}\right) \tag{15}$$

where, C_o is observed values from regression equation and C_P is predicted. These parameters were used to further assess the predictability. The values of these statistical parameters are presented in Table 3.

Statistical Parameter	Complete Dataset	Application		Post-Application
		Low Wind	Very High Wind	High Wind
NMSE	0.17	0.31	0.017	0.21
Fractional Bias	0.23	0.41	0.09	0.21
R	0.94	0.96	0.89	0.71
MG	0.78	0.90	0.65	0.80

Table 3. Shear Layer Model Performance Using Predicted and Measured Concentrations

For a "perfect" ideal model the fractional bias and the normalized mean square error are equal to zero. The ideal values for a geometric mean bias and the correlation coefficient should be 1. As expected in the real life, the shear layer model is not a perfect model. However, the acceptable range for NMSE and FB for an air quality model suggested by Kumar et al. (1993) is given as, NMSE $\leq$ 0.5 and -0.5 $\leq$ FB $\leq$ 0.5. The values of NMSE and FB for shear layer model in all wind conditions were within acceptable limits.

The geometric mean bias is a function of a logarithmic mean of the predicted and observed data. Geometric mean bias values of 0.5-2.0 can be thought as "factor of two" over predictions and under predictions in the mean respectively (Hanna et al., 1993). Thus the geometric mean range for the acceptable model is given as 0.5 $\leq$ MG $\leq$ 2.0. When a data set contains pairs of data 10 or less, then the logarithmic forms are appropriate, so that the under predictions and the over predictions receive equal weight. The values of MG for each condition are better representation of the behavior of a model to assess whether a model is over predicting or under predicting in a particular situation. From Table 5 it was observed that the shear layer model over predicts the concentrations under almost all the conditions. This may be due to the factors such as the use of concentrations measured at 1.5 m as ground level concentrations, the concept of eddy diffusivity for atmospheric turbulence in the new model, and the assumptions made for other model inputs. It was also observed that during the low wind conditions the predictions were closer to reality (MG=0.90) than during other wind conditions.

5.2 Model evaluation using literature data

To evaluate the model based on the literature data, an evaluation case was developed based on Brooks et al. (2005) study. The paper gives a regression equation based on the data collected downwind of the application site. For this evaluation purpose a constant emission rate of 4.13 particles/ m^2/sec as given in the paper was used. Wind velocity was 2.29 m/s at 10 m height. Based on the atmospheric conditions described in the literature, the slightly stable to near neutral stability condition was assumed for the simulation. The input values for the stability parameters used for shear layer model were used from Table 2.

The predicted concentrations were plotted along with the concentrations obtained from the regression equation for various downwind distances (See Figure 2). The comparison of predicted concentration with the observed concentration from regression equation was plotted (See Figure 3). It was observed from the figures that shear layer model under predicts the concentration for shorter downwind distance ($x < 15$ m) closer to the field, but for the higher downwind distances ($x > 20$ m) the model over predicts the concentrations. As a result, the shear layer model, again, was observed to over predict the downwind concentrations.

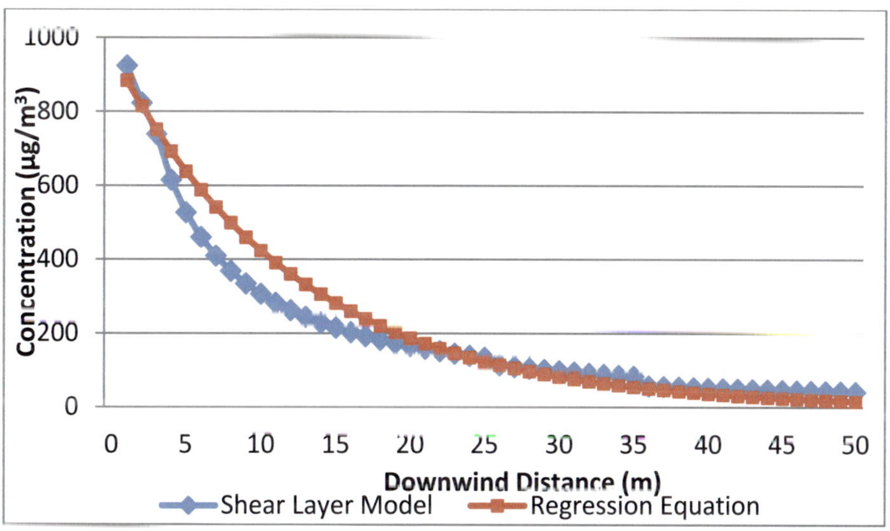

Fig. 2. Comparison of Concentrations predicted using the Shear Layer Model and Regression Equation by Brooks et al. (2005)

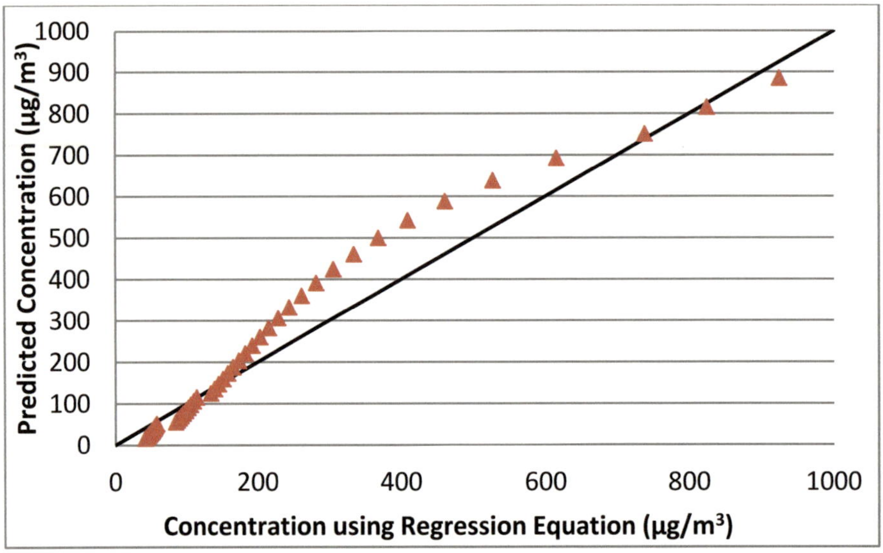

Fig. 3. Concentration using Regression Equation vs. Predicted Concentration from the Shear Layer Model

Again, performance measures were calculated from the modeled and the observed concentrations. The statistical parameter NMSE, FB, correlation coefficient, and geometric mean (MG) were calculated using the previously stated equations (See Table 4). It was determined from these performance measures that even though the shear layer model was not a perfect model, the parameters were within the acceptable range for a good fit model. The geometric mean bias indicates that the shear layer model over predicts the downwind concentrations for this data set.

As seen from the model evaluation figures and statistical evaluation, the model produced consistently good performance in simulating the downwind concentration from the application and the post-application. The model performance was also good in varying wind conditions. From the performance measures it was determined that the model over predicts the concentrations in most cases. This evaluation was performed using the limited measured and literature data available at the time of the research.

Statistical Parameter	Value
NMSE	0.14
Fractional Bias	-0.1
R	0.95
MG	0.89

Table 4. Statistical Parameter Calculated for Evaluation of the Shear Layer Model based on Regression Equation

6. Conclusion

The objective of this chapter was to develop and evaluate a dispersion model for particulate matter associated with biosolids application on a farm field. The following observations were made:

1. An analytical solution to convective-diffusion equation (the shear layer model) to incorporate wind shear near the ground was presented to predict the downwind concentration of total particulate matter. The shear layer model was evaluated using limited field study data. The model was observed to over predict the concentration for the low wind conditions during the application. For the high wind conditions during the post-application, the model was under predicting the concentration. The statistical parameters revealed that the shear layer model is a good fit to the measured data.
2. The concentrations predicted were compared to the observed regression concentrations from the literature. The results showed that shear layer model under predicts at the lower downwind distances whereas it over predicts at higher downwind distances. Again the statistical parameters revealed shear layer model to fit the literature data.

A generic screening model was derived, and can be used to predict the downwind concentrations of particulate matter emitted from the land application of biosolids. It was observed that the model over predicts the downwind concentrations in unstable conditions. Future work should focus on performing field studies to collect data under different atmospheric conditions.

7. Acknowledgement

The authors would like to thank Dr. Farhang Akbar and Ms. April Ames at The University of Toledo for the particulate data collection during the field study in the summer 2009. The funding provided by the U.S. Department of Agriculture, USDA-2008-38898-19239, and USDA-2009-38898-20002 is gratefully acknowledged. The views expressed in this paper are those of the authors.

8. References

Akbar-Khanzadeh A., Ames A., Bisesi M., Milz M., Kumar A. (In review), Particulate Matter Characteristics in Relation to Environmental Factors in a Biosolids-applied Farm Field

Brooks J.P., Tanner B.D., Gebra C.P., Haas C.N., Pepper I.L. (2005), Estimation of Bioaerosols Risk of Infection to Residents Adjacent to a Land Applied Biosolids Site using an Empirically Derived Transport Model, *Journal of Applied Microbiology*, 98,397-405

Davis, G. (2002), Western Australian Guidelines for Direct Land Application of Biosolids and Biosolids product, Department of Environmental Protection, Perth, WA, Government of Western Australia, 110 pp

Dowd S.E., Gebra C.P., Pepper I.L., Pillai S.D. (2000), Bioaerosols Transport and Risk Assessment in Relation to Biosolid Placement, *Journal of Environmental Quality*, 29, 343-348

Gudivaka V., Kumar A. (1990), An Evaluation of Four Box Models for Instantaneous Dense-Gas Releases, *Journal of Hazardous Material*, Vol. 25, pp. 237-255

Hanna S.R., Chang J.C., Strimaitis D.G. (1993), Hazardous Gas Model Evaluation with Field Observations, *Atmospheric Environment*, 27A (15), 2265 2285

Kumar A. (1977), Pollutant Dispersion in the Planetary Boundary Layer, *PhD Dissertation*, The University of Waterloo, Ontario, Canada, 182 pp

Kumar A., Luo J., Bennett G.(1993), Statistical Evaluation of Lower Flammability Distance (LFD) using Four Hazardous Release Models, *Process Safety Progress*, 12(1), pp. 1-11

Kumar A., Dixit S., Varadarajan C., Vijayan A., Masuraha A. (2006), Evaluation of the AERMOD Dispersion Model as a Function of Atmospheric Stability for an Urban Area, *Environmental Progress,* 25(2), 141-151

Kumar A., Bhat A. (2008), Development of a Spreadsheet for the Study of Air Impact due to Releases of bioaerosols, *Environmental Progress,* 27(1), 15-20

Paez-Rubio T., Xin Hua., Anderson J., Peccia J. (2006), Particulate Matter Composition and Emission Rates from the Disk Incorporation of Class B Biosolids into Soil, *Atmospheric Environment,* 40, 7034-7045

Riswadkar R.M., Kumar A. (1994), Evaluation of the ISC Short Term Model in a Large-Scale Multiple Source Region for Different Stability Classes, *Env. Monitoring and Assessment,* 1-14,

Sutton O.G. (1953), Diffusion and Evaporation, *Micrometeorology,* McGraw-Hill Book Company, NY, 273-323

Taha MPM., Pollard SJT., Sarkar U., Longhurst P. (2005), Estimating Fugitive Bioaerosols Releases from Static Compost Windrows: Feasibility of Portable Wind Tunnel Approach, *Waste Management,* 25, 445-450

3

Air Pollution: A Case Study of Ilorin and Lagos Outdoor Air

A.M.O. Abdul Raheem and F.A. Adekola
Department of Chemistry, University of Ilorin, Ilorin, Nigeria

1. Introduction

Air pollutants are continuously released from numerous sources into the atmosphere. Several studies have been carried out on the quantification of pollutants and analyzing their consequences on public health. It has been estimated that each year between 250 and 300 million tons of air pollutants enter the atmosphere above the United States of American [Dara, 2004; Onianwa, 2001; Stephen and Spencer, 1992]. Tropospheric pollution causes degradation of crops, forests, aquatic systems, structural materials, and human health. It was reported recently, that NO_x air pollution is becoming a far reaching threat to USA National Parks and Wilderness Areas as these areas are suffering from harmful effects of oxides of nitrogen pollution [EDFS, 2003]. It has also been confirmed that NO_x contributes to ground - level ozone (smog) pollution which can cause serious respiratory problems, especially young children and the elderly, as well as healthy adults that are active outdoors. Furthermore, the same report confirmed worsening ozone concentration in nearly all the national parks over the last ten years [EDFS, 2003]. Towards this, an assessment of new vehicles emission certification standards was carried out in metropolitan area of Mexico city and the results show that light duty gasoline vehicles account for most carbon (II) oxide and NO_x emissions [Schifter *et al*, 2006]. The European Environmental Agency also reported very recently that more than 95% contribution to nitrogen oxides emission to the air comes from fuel combustion processes from road transport, power plants and industrial boilers [EEA, 2006]. There is reported evidence of average chronic damage to the human lung from prolonged ozone exposure [EEA, 2006]. Sulphur in coal, oil and minerals are the main source of the Sulphur (IV) oxide in the atmosphere. Moreover, peak concentrations above European Union limit still occur, especially close to point sources in the cities. Asian cities have some of the highest levels of air pollution in the world. In Asia, hundreds of thousands of people in urban areas get sick just by breathing the air that surrounds them. However, the WHO 2006 estimates that dirty air kills more than half a million people in Asia each year of which burden falls heaviest on the poor as reported by Ogawa, 2006. The worsening of the situation has been attributed to cumulative effects of rapid population growth, industrialization and increased use of vehicles. The ozone primary tropospheric pollutants (SO_2, NO_x, HC_s, and CO) often react in the atmosphere to form secondary pollutants which are acidic compounds (H_2CO_3, H_2SO_4 and HNO_3) and photochemical oxidants. Environmental damage frequently results from several primary and secondary pollutants acting in concert rather than from a single pollutant. Tropospheric oxidants such as ozone, PBN, PAN illustrate the complexity of atmospheric chemistry and processes. They help to form acidic compounds thereby contributing to green house warming and hence, damage to human health, animal, plant life and materials [USEPA, 1998; Dara, 2004]. Significant changes in stratospheric ozone, high above the troposphere, can affect tropospheric oxidants level [USEPA, 1998]. If increased UV-B radiation penetrates a depleted ozone shield, the photochemical formation of ground level oxidants may be enhanced [Stoker and Seager, 1972]. Green house warming could amplify this effect: A study carried out in three U.S. cities; Nashville, Philadelphia, and Los Angeles showed that a large depletion of stratospheric ozone, coupled with the green house warming, could increase smog formation by as much as 50% [Adelman, 1987]. The study also showed that NO_2 concentration might increase more than ten folds. There is however progress towards the reduction of anthropogenic emissions of NO_x, CO, volatile organic compounds in Europe and North American [Jonson *et al*, 2001]. However, the concentration of air pollutants emitted into the atmosphere is on the increase in the Southeast Asia and other parts of the World [Jonson *et al*, 2001]. It is therefore expected that the emissions from Africa and other parts of the Worlds that are yet to take strict and effective controlling measures on emissions will influence the free tropospheric levels in most of the Northern Hemisphere.

During a five-day period marked by temperature inversion and fog in London in 1952, between 3,500 and 4,000 deaths in excess of normal occurred with 1.3 ppm SO_2 level recorded [O'Neill, 1993; ACGIH, 1991]. SO_2 is oxidized to SO_3 in the atmospheric air by photolytic and catalytic processes involving ozone, oxides of nitrogen (NOx), and hydrocarbon (HC), giving rise to the formation of photochemical smog [Dara, 2004].

In contrast to NO_2, SO_2 is deleterious to plant life [Haagen-Smith, 1952; Molski and Dmuchowski, 1986]. Air pollution causes the decline in Eastern Europe [Nihlgard, 1985; Schutt and Cowling, 1985]. Area with extensive large-scale forest decline correlate with the areas where SO_2 concentration were elevated [ECE, 1984; Molski and Dmuchowski, 1986], therefore If the concentration of SO_2 is higher along with other gaseous pollutants (NO_x, O_3, HC, organic and inorganic peroxides etc) in the troposphere and continues to accumulate over time, the overall concentration can have a negative effect on health, vegetation and structures [Abdul Raheem, 2007; Abdul Raheem et al., 2009[a]].

Surface concentration of ozone has been reported to be on the increase in the last decades in northern hemisphere [Vautard et al., 2007] and in Southeast Asia [Jaffe, 1999]. A rise of about 26% in the ambient concentration of ozone has been reported in Taiwan between 1994 and 2003 [Chou et al., 2006]. High levels of ambient air ozone can cause serious damage to health. The health hazards include shortness of breath, nausea, eye and throat irritation, and lung damage [Menezes and Shively, 2001]. Identification of air pollution source characteristics is an important step in the development of regional air quality control strategies. Receptor modeling, using measurements of pollutant concentrations at one or more sample sites, is often a reliable way to provide information regarding source regions for air pollution [Watson, 1984]. One of such receptor-modeling technique is principal component analysis (PCA) [Einax and Geiss, 1997; Jackson, 1991; Norman, 1987]. This is often combined with multiple linear regression (MLR), principal component regression (PCR), and partial least square regression, which have been demonstrated as powerful tools for handling several environmental problems, especially source apportioning [Otto, 1999; Timm, 1985; Vogt, 1989].

In recent years, certain statistical techniques that incorporate the influence of meteorological variables have been applied to asses the trend in ozone levels in the ambient air [Bakken et al., 1997]. One common approach is the use of a parametric regression model to link some characteristics of ozone, such as the mean level of ozone to meteorological variables. Other scientists have equally used PCA to pattern the spatial and temporal variations of ozone and to identify the important factors influencing ozone concentration [Klaus et al., 2001; Lengyel et al., 2004; Pissimanis et al., 2000]. Different subregions have, however, been identified where ozone concentration exhibited characteristic spatial and temporal patterns based on the differences arising from the interaction of their respective meteorological conditions with anthropogenic effects [Alvarez et al., 2000]. More specifically, Bloomfield et al., (1996) established a nonlinear regression model for hourly ozone data in the Chicago area in which meteorological variables, seasonality, and a trend term were all implicated. Cox and Chu (1992), on their part, proposed a model for the daily maxima of hourly ozone concentrations based on the Weibull distribution in which the scale parameter is allowed to vary as a function of meteorological conditions. On the other hand, Menezes and Shively (2001) used a multivariate approach to estimate the long-term trend in the extreme values of tropospheric ozone in Houston and Texas. They found that there is a downward trend in the probability of an exceedance followed by a relative flat trend. Shively and Sager (1999) extended the work of Cox and Chu (1992) as well as Bloomfield et al., (1996) by using nonparametric regression models to model ozone. The use of multivariate methods was further supported by Lengyel et al., (2004) who analyzed air quality data of which the hidden structure was uncovered by factor analysis and modeled ozone concentration using MLR, PLS, and PCR. While PCA is a very useful tool for selection of properties and different qualities processes leading to a linear model of the data, MLR and PCR or PLS can predict ozone concentration with an error below 2, 5, and 1ppb levels, respectively.
The aim of the present investigation was therefore to analyze environmental data gathered on the daily monitoring of ambient ozone, oxides of nitrogen, and sulfur (IV) oxide at five monitoring sites in Lagos and four monitoring sites in Ilorin, Nigeria. The data were collected between early morning and late evening of the day, and they covered both dry and wet seasons from years 2003 to 2006. The study also included establishment of prediction models on the influence of meteorological parameters on the seasonal variation of concentration of gaseous pollutants in the two Nigerian cities and comparison of measured and modeled concentration values of ozone [Abdul Raheem et al., 2009b].

2. Theoretical background

2.1 Air pollution

Air is all around us, odourless, colourless and essential to all life on earth as it acts as a gaseous blanket, protecting the earth from dangerous cosmic radiation from outer space. It helps in sustaining life on earth by screening the dangerous ultraviolet (UV) radiations (< 300nm) from the sun and transmitting only radiations in the range 300nm to 2500nm, comprising of near UV, visible and near infrared (IR) radiations and radio waves (0.01 to 1 x 10^5nm) [Smart, 1998]. The atmosphere also plays a vital role in maintaining the heat balance on the earth by absorbing the IR radiation received from the sun and re-emitted by the earth. In fact, it is this phenomenon, called "the greenhouse effect", which keeps the earth warm enough to sustain life on the earth. Yet, the air is actually a combination of gaseous elements that have a remarkable uniformity in terms of their contribution to the totality of life. Thus, oxygen supports life on earth; nitrogen is an essential macro - nutrient for plants; and carbon (IV) oxide is essential for photosynthetic activity of plants. Moreover, atmosphere is a carrier of water from the ocean to land, which is so vital for the hydrological cycle. Any major disturbance in the composition of the atmosphere resulting from anthropogenic activities may lead to disastrous consequences or may even endanger the survival of life on earth [Dara, 2004]. The constituent elements are primarily nitrogen and oxygen, with a small amount of argon. Below 100km, the three main gaseous elements, which account for about 99.9% of the total atmosphere, are N_2, O_2 and Ar and they have concentration by volume of 78%, 21%, and 0.93% of respectively [Stanley, 1975]. The presence of trace amounts of other gases would account for the remaining 0.07%. These remaining trace gases exist in small quantities and they are measured in terms of a mixing ratio. This ratio is defined as the number of molecules of the trace gas divided by the total number of molecules present in the volumes sampled. For example, ozone (O_3), carbon (IV) oxide (CO_2), oxides of nitrogen (NO_2 + NO) as NO_x and chloro fluoro carbons (CFCs) are measured in parts per million by volume (ppmv), parts per billion by volume (ppbv) as well as microgram per cubic meter (μgm^{-3}), [Dale, 1976]

Air pollution has been defined [World Bank, 1978] as 'the presence in the outdoor atmosphere of one or more contaminants such as dust, fumes, gas, mist, odour, smoke, or vapour in such quantity, characteristics and duration as to make them actually or potentially injurious to human, plant or animal life, or property, or which unreasonably interferes with the comfortable enjoyment of life and property'. Pollution on the whole is caused principally by human activities, though it can also be a natural process. Air pollution arises from people's economic and domestic activities like modern agriculture which requires agrochemicals. Industrial activities are responsible for wide range of pollution. Thermal power station, burning fossil fuel and moving vehicles emit harmful pollutants like sulphur (IV) oxide, nitrogen (II) oxide and carbon (IV) oxide. Some of these emitted gases have been responsible for acid - rain, global warming and malfunctioning of human / animal's haemoglobin [Stanley, 1975]. Other causes arising from human activities include inappropriate solid waste disposal, gas flaring and oil exploration. Air pollution can also arise from natural causes such as volcanic eruption, whirlwind, earthquake, decay of vegetation, pollen dispersal, as well as forest fire ignition by lightning.

2.2 Pollutants

These are substances introduced into the environment in an amount sufficient to cause adverse measurable effects on human beings, animals, plant, vegetation or materials. Pollutants are referred to as primary pollutants, if they exert the harmful effects in the original form in which they enter the atmosphere e.g. CO, NO_x, HC_s, SO_x, particulate matter and so on. On the other hand, secondary pollutants are products of chemical reactions, among primary pollutants are ozone, hydrogen peroxide, peroxyacetylnitate (PAN) and peroxybenzoyl nitrate (PBN). Classification of pollutants can also be according to chemical compositions i.e. organic or inorganic pollutants or according to the state of matter i.e. gaseous or particulate pollutants. Air pollution is basically made up of three components and these are source of pollutants, the transporting medium, which is air and target or receptor which could be man, animal, plant and structural facility.

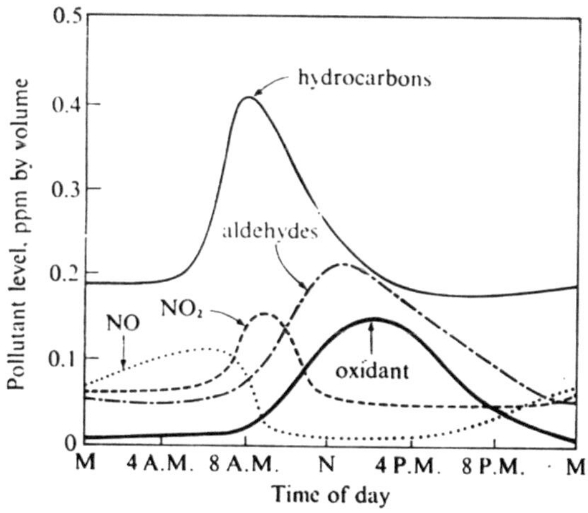

Fig. 1. Generalized Plot of Atmospheric Concentrations of Species Involved In Smog Formation As A Function of Time of The Day

3. Atmospheric photochemical reactioN

The various chemical and photochemical reactions taking place in the atmosphere mostly depend upon the temperature, composition, humidity and the intensity of sun light. Thus the ultimate fate of chemical species in the atmosphere depends upon these parameters. Photochemical reactions take place in the atmosphere by the absorption of solar radiations in the UV region. Absorption of photons by chemical species gives rise to electronically excited molecules. These reactions are not possible under normal laboratory conditions except at higher temperature and in the presence of chemical catalysts [Hansen *et al*, 1986]. The electronically excited molecules spontaneouslyundergo any one or combination of the following transformations: Reaction with other molecules on collision; Polymerization; Internal rearrangement; Dissociation; De - excitation by fluorescence or De - activation to

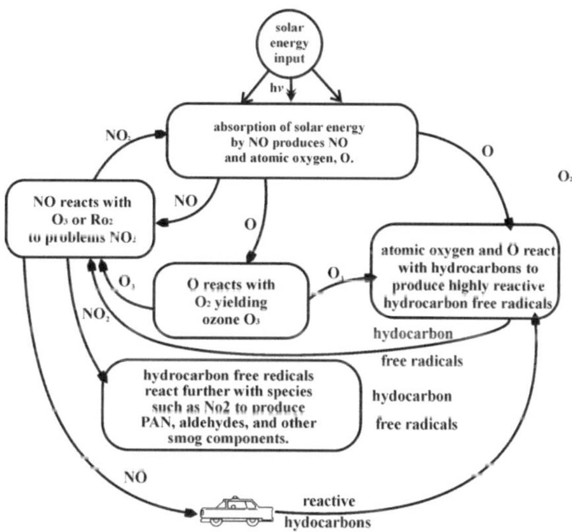

Fig. 2. Generalized Scheme For The Formation Of Photochemical Smog

return to the original state [Dara, 2004]. Any of these transformation pathways may serve as an initiating chemical step or a primary process. The three steps involved in an overall photochemical reaction are Absorption of radiation, Primary reactions and Secondary reactions.

Smoggy atmosphere show characteristics variations with time of the day in levels of NO, NO_2, hydrocarbons, aldehydes and oxidants. A generalized plot showing these variations is shown in Figure 1. This shows that shortly after dawn the level of NO in the atmosphere decreases markedly, a decrease which is accompanied by a peak in the concentration of NO_2. During the mid - day the levels of aldehydes and oxidants become relatively high, however, the concentration of total hydrocarbons in the atmosphere peaks sharply in the morning, then decreases during the remaining daylight hours. The variations in species concentration shown in the above Figure 1 may be explained by a generalized reaction scheme in Figure 2. This is based on the photochemically initiated reactions which occur in an atmosphere containing oxides of nitrogen, reactive hydrocarbons, and oxygen. The various chemical species that can undergo photo - chemical reactions in the atmosphere include NO_2, SO_2, HNO_3, N_2, ketones, H_2O_2, organic peroxides and several other organic compounds and aerosols; the time - variations of which are explained in a group of overall reactions first proposed by Friedlander nd Seinfeld (1974)

* Primary photochemical reaction:

$$NO_2 + h\nu \rightarrow NO + O \quad (1)$$

* Reactions involving oxygen species

$$O_2 + O + M \rightarrow O_3 + M \quad (2)$$

(M is an energy absorbing third body)

$$O_3 + NO \rightarrow NO_2 + O \quad (3)$$

* Production of organic free radicals from hydrocarbons,

$$RHO + RH \rightarrow R\cdot + \text{other products} \quad (4)$$

$$O_3 + RH \rightarrow R\cdot + \text{other products} \quad (5)$$

R· is a free radical, which may or may not contain oxygen

* Chain propagation, branching and termination

$$NO + R\cdot \rightarrow NO_2 + R^1 \quad (6)$$

In this case R· contains oxygen and oxidizes NO. It is one of many chain propagation reactions, some of which involve NO

$$NO_2 + R\cdot \rightarrow \text{products (e.g. PAN)} \quad (7)$$

A number of specific reactions are involved in the above overall scheme for the formation of photochemical smog, which is smoke and fog [Thomas *et al*, 1974]. The formation of atomic oxygen by the primary photochemical reaction:

$$NO_2 + h\upsilon \rightarrow NO + O$$

This leads to several reactions involving oxygen and nitrogen oxide species. Examples of such reactions are given below:

$$O + O_2 + M \rightarrow O_3 + M \quad (8)$$

$$O + NO + M \rightarrow NO_2 + M \quad (9)$$

$$O + NO_2 \rightarrow NO + O_2 \quad (10)$$

$$O_3 + NO \rightarrow NO_2 + O_2 \quad (11)$$

$$O + NO_2 + M \rightarrow NO_3 + M \quad (12)$$

$$O_3 + NO_2 \rightarrow NO_3 + O_2 \quad (13)$$

Pollutant Source	CO	NOx	HC	SOx	Particulate	Total
Transportation	111.0	11.7	19.5	1,0	0.7	143.9
Fuel Combustion (Stationary Source)	0.8	10.0	0.6	26.5	6.8	44,7
Industrial Processes	11.4	0.2	5.5	6.0	13.1	36.2.
Solid wastes disposal	7,2	0.4	2.0	0.1	1.4	11.1
Miscellaneous	16.8	0.4	7.1	0.3	3.4	28.0
Total	147.2	22.7	34.7	33.9	25.4	263.9

[Adapted from Thomas *et al*, 1974]

Table 1. Estimates of USA Primary Pollutants Sources in million tons per year (Million Tons/Year)

4. Methodology

The determinations of the concentrations of total oxidants (undertaken as O_3), NO_X and SO_2 in the ambient air were carried out between 2003 and 2006 to cover the two seasons; dry season (November to April) and rainy season (May to October) using standard methods. Total oxidants were determined by buffered potassium iodide solution method proposed by Byers and Saltzman (1958). Determination of oxides of nitrogen concentrations were done using the Intersociety Committee Method of Analysis (1972) which is based on the Griess-Saltzman (1954) colorimetric, azo dye forming reagent while oxides of sulphur were determined by conductivity measurements as proposed by Stanley (1975). The application of these techniques to monitoring required the preparation of the following range of reagents as sampling and absorbing solutions as well as preparation of calibration curves. All the chemicals used in the preparation of sampling and absorbing solutions were of analytical grade. The reagents had been prepared and seen stable for months prior to their use. The validation of all procedures was further confirmed using LaMotte standard air sampler.

4.1 Ozone

The absorbing solution used for trapping ozone was 1% KI buffered at 6.8 ± 0.2. The standard solution and calibration curve were prepared as follows: 4.09ml of standard 0.001M iodine solution was taken and diluted to 100ml with distilled water. 10ml of this solution was taken and further diluted to 100ml with absorbing reagent and this was labelled solution A. 1ml of A was further diluted to 100ml using absorbing reagent, which gives solution B. This provides a calibrated iodine solution equivalent to 1.92μg of ozone per ml [Stanley, 1975 and Dara, 2004] as ozone reacts with iodine ion in neutral buffer solution according to the following equation:

$$O_3 + 3I^- + H_2O \longrightarrow I_3^- + 2\,OH^- + O_2 \qquad (14)$$

The tri - iodide ion liberated has an intense yellow colour. The standard solution was always prepared freshly when needed. A series of standard solutions prepared from above were used to obtain calibration curve. The absorbance measurement was carried out at 352nm. The calibration curve is shown in Figure 3 and it has regression value of 0.9972

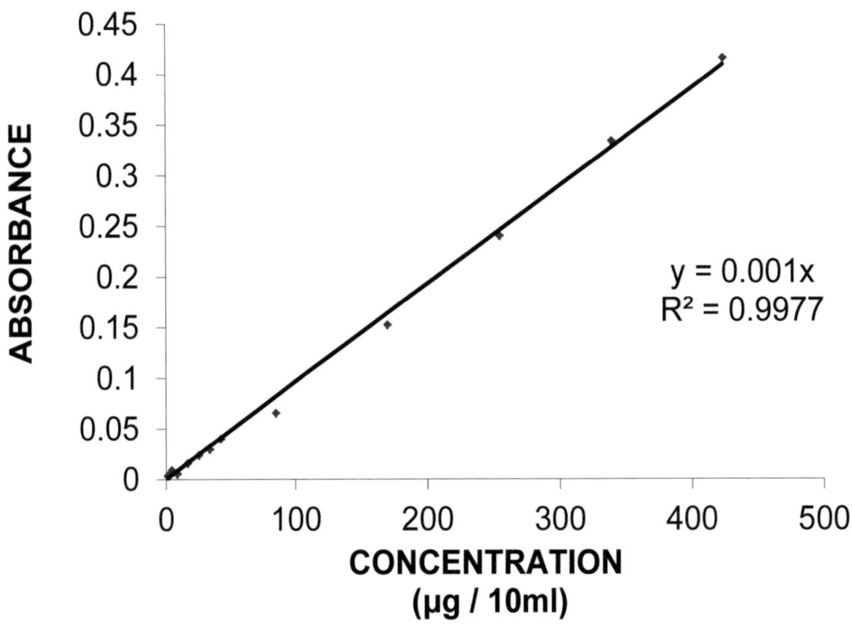

Fig. 3. Total oxidants calibration curve

4.1.1 Sampling procedure

The set-up of the high volume sampler is given in Figure 4. To follow strict guidelines needed when monitoring for criteria pollutants as discussed earlier, our equipment is validated with LaMotte air sampler (Figure 5) purchased from LaMotte & Company, USA, for capability, repeatability and reliability needed to collect accurate data, and operation of the equipment within our established methods. The impinger used is a big boiling tube of capacity 250cm³ with 42mm diameter. Teflon tubing's used as delivery tubes along with

Fig. 4. Picture of The Sampling train

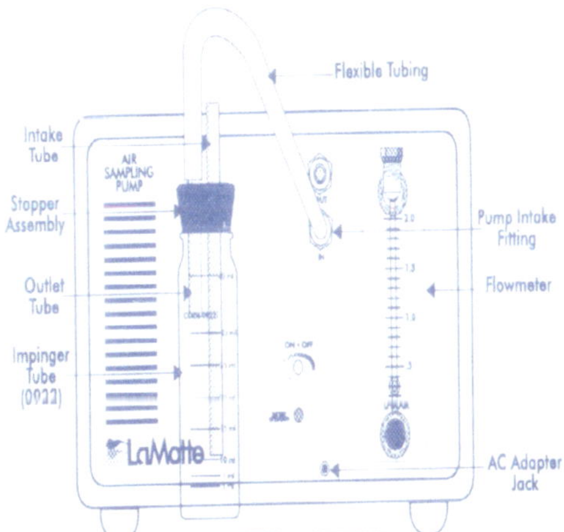

Fig. 5. Lamotte Air- Sampler

glass tubing that serves as inlet for the ambient air sampled. Silicone grease was used to make the set up airtight by it application to all necessary joints. The air was sampled at the rate of 1 dm^3 min $^{-1}$ with absorbing solution fixed at 30 cm^3 as found appropriate and used [Abdul Raheem et al., 2009c], after each sampling for one hour, the impinger was carefully removed and the sample transferred quantitatively into the sample bottle for analysis. The impinger was wrapped with aluminum foil to avoid sunrays' interference. In order to determine pollutant variability over daytime periods, air samples were taken at each location over ten defined sixty - minute periods for any sampling day. The ten sampling periods were spaced equally between approximately 6 am and 6 pm to reflect morning, afternoon and evening. Minimum of 30 samples were collected at each site per week for the pollutants monitored. The time of sixty minutes was found to be optima from the preliminary investigations for the quantitative sampling of these toxic gases within the

environment [Abdul Raheem et al., 2009c]. All samplings reported were carried out in triplicates. As the road traffic is the common source of pollution cutting across all sites classes, the average traffic volume were determined for all sampling zones. Traffic count was manually done, counting the vehicles passing on the road for 10 minutes in every hour from which hourly traffic was calculated [Abam and Unachukwu, 2009].
The daily minimum and maximum temperatures were between 23°C and 36.5°C, throughout the sampling period.

4.1.2 Analysis

A freshly prepared absorbing solution serves as sample reference or blank solution in order to take care of any impurities during preparation. Absorbance of samples for total oxidants was measured at 352 nm with UV / Visible spectrophotometer. The concentration was read out in µg / 10ml from the reference plot of which one of the examples is shown in Figure 3. The concentrations were converted to μgm^{-3} or ppm or ppb using appropriate conversion factor.

4.1.3 Calculation [Vowels and Connell, 1980]

$$\text{Oxidant}\left(\mu\text{gm}^{-3}\right) = \frac{\text{total } \mu\text{g O}_3\text{per 10ml of absorbing reagent}}{\text{Volume of air sampled in cubic metres}} \tag{15}$$

$$O_X = \frac{M_{O3}}{V} \tag{16}$$

$$\mu gm^{-3} = \frac{ppb \times molar\, mass}{24.45} \times 1000 \tag{17}$$

for 1 μgm^{-3} of ozone, the ppb value will be

$$\frac{1 \times 24.45}{48 \times 1000} = 0.51 ppb \tag{18}$$

QUANTITY	CONTENTS	CODE
2 × 120 Ml	Total oxidants reagent #1	7740-J
30Ml	Total oxidants reagent #2	7741-G
30Ml	Total oxidants reagent #3	7742-G
3	Test Tubes, 5mL, w/ caps	0230
1	Total oxidants in Air Comparator	7739

Table 2. LaMotte total oxidants in air test kit code 7738

4.1.4 La Motte total oxidants sampling procedure

10 mL of reagent #1 was put into impinging tube, followed by 2 drops of reagent #2 added and swirled to mix then 2 drops of reagent #3 added and also swirled to mix. The impinging apparatus was connected to intake of the sampling pump as shown in Figure 5 such that the long tube was immersed in the absorbing solution. The impinging tube was covered with foil to protect it from light while sampling. The flow meter of sampling apparatus was adjusted to collect air at 1.0 Lm^{-1} rate. The sampling continued until 15 minutes when a measurable pink colour developed. The impinging tube was disconnected from the pumping apparatus and the contents poured into a clean test tube (0230). The test tube was later inserted into the total oxidants in air comparator (7739) and the sample colour was matched with an index value. The index value was recorded and the calibration chat was used to convert the index readings into concentration of the pollutant in the atmosphere in parts per million.

Time(min)	1	2	3	4	5	6	7	8
5	0.14	0.36	0.72	1.08	1.44	2.88	4.32	5.76
10	0.07	0.18	0.36	0.54	0.72	1.44	2.16	2.88
15	0.05	0.12	0.24	0.36	0.48	0.96	1.44	1.72

** Values in ppm

Table 3. Total oxidants in air calibration chart** [LaMotte 6.05] Comparator index number

4.2 Oxides of nitrogen (NO_x)

The absorbing solution used for trapping NO_x was Saltzman solution which is an azo dye forming reagent .

The standard solution and calibration curve were prepared as follows: 2.16g of sodium di - oxo nitrate (III), $NaNO_2$ was dissolved in 1000 cm^3 volumetric flask and the solution labeled A. 1ml of solution A was measured out into 100 ml volumetric flask and the solution made up to the mark. This solution of concentration 0.0216 gL^{-1} was labeled B. 1ml of B was added to 100 ml volumetric flask and distilled water added to the mark to give 0.000216 gL^{-1} solution C. 10ml of solution C was added to 100ml volumetric flask and filled to the mark with distilled water to give solution of concentration 0.0000216 gL^{-1} labeled D. Further dilutions of the last two solutions C and D were used for calibration plot. As the standardization was based on the empirical observation that 0.72 mole of $NaNO_2$ produces the same colour as 1mole of NO_2 [Hesketh, 1972]. In other words, 1ml of the 0.000216 gL^{-1} working standard which contains 0.216μg of $NaNO_2$ should be equivalent to 0.2 μg of NO_2. Series of standard solutions prepared in 10 ml volumetric flasks from solutions C and D above were allowed to stay for 15 minutes for colour development and the spectra run at 550 nm to obtain a set of absorbance value which were recorded against known concentrations. The formation of red azo dye of which the absorbance is picked at 550 nm can be explained according to the equation in Figure 6 However, a plot of absorbance against concentration in μg / 10 ml was made, a straight line graph obtained with regression value of 0.9962 as shown in Figure 7

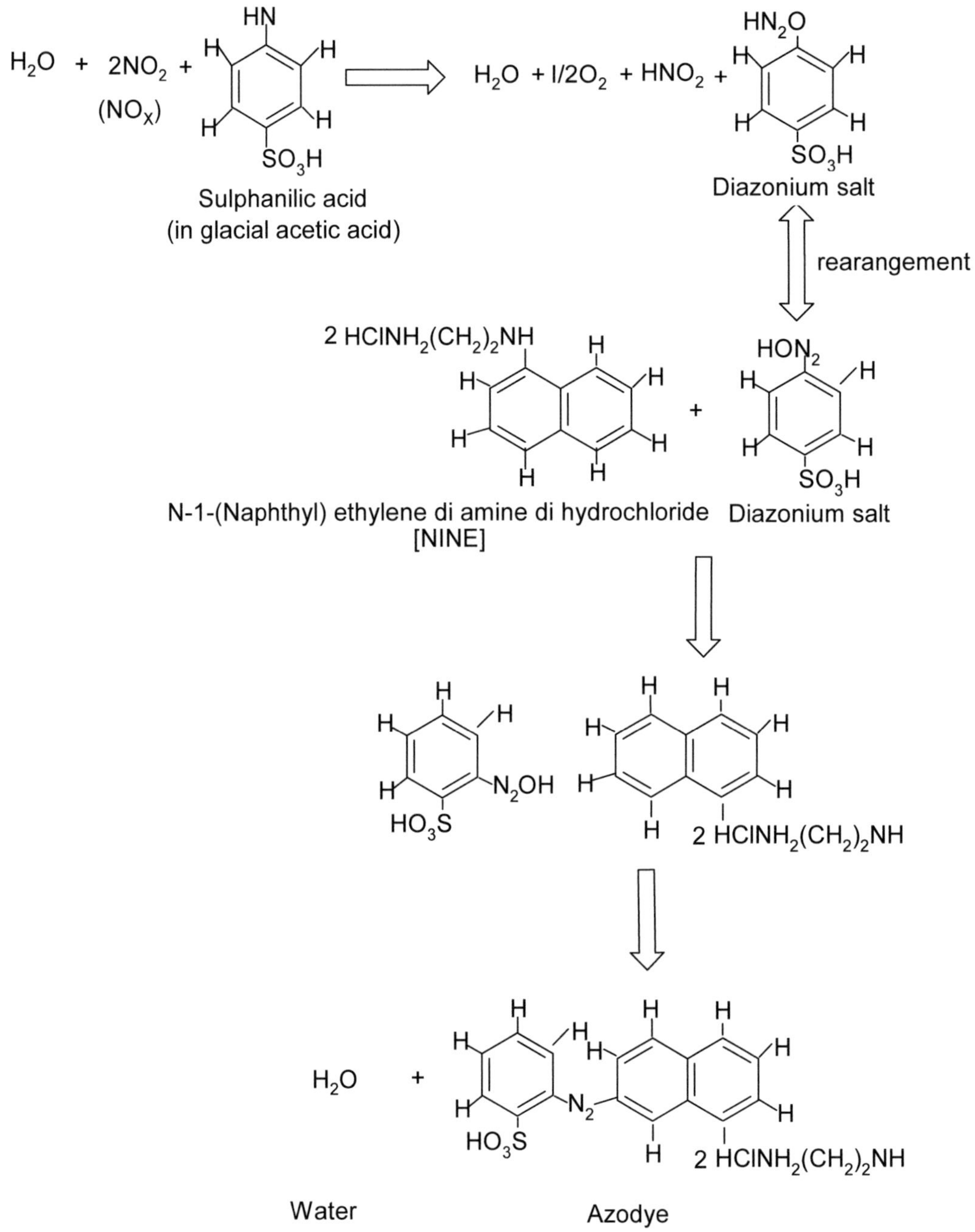

Fig. 6. Equation showing the formation of azo dye

4.2.1 Sampling procedure
The procedure for sampling is as given above

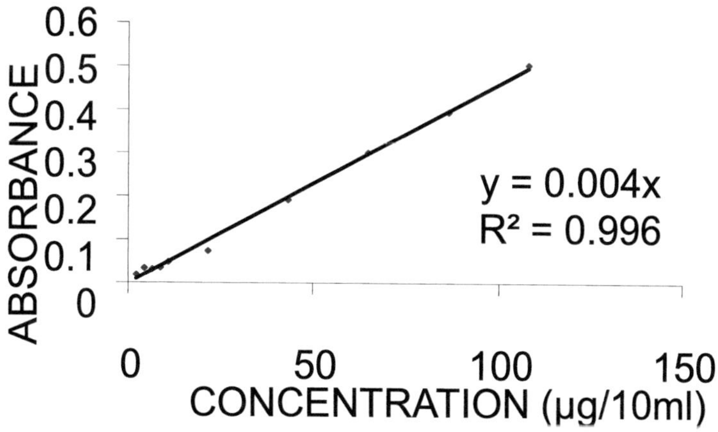

Fig. 7. Oxides of nitrogen (no_x) calibration curve

4.2.2 Analysis
The absorbing solution serves as sample reference or blank solution in order to take care of any impurities during preparation. Absorbance of samples for oxides of nitrogen was measured at 550nm with UV / Visible spectrophotometer. The concentration was read out in µg / 10ml from the reference plot of which an example is shown in figure 7. The concentrations were converted to μgm^{-3} or ppm or ppb of which the conversion factors are explained hereafter.

4.2.3 Calculation [Vowels and Connell, 1980]

$$NO_x\left(\mu gm^{-3}\right) = \frac{\text{total } \mu g \text{ } NO_x \text{ per 10ml of absorbing reagent}}{\text{Volume of air sampled in cubic metres}} \quad (19)$$

$$NO_X - \frac{M_{NO_X}}{V} \quad (20)$$

$$\mu gm^{-3} = \frac{ppb \times molar\ mass \times 1000}{24.45} \quad (21)$$

for 1 μgm^{-3} of NO_x as NO_2, the ppb value will be

$$\frac{1 \times 24.45}{46 \times 1000} = 0.53 ppb \quad (22)$$

QUANTITY	CONTENTS	CODE
2 × 120 mL	Nitrogen (Iv) oxide reagent #1 Absorbing solution	7684-J
30 mL	Nitrogen (Iv) oxide reagent #2	7685-G
10g	Nitrogen (Iv) oxide reagent #3 powder	7688-D
2	Test tubes,10mL, glass, w/caps	0822
1	Spoon, 0.005g, plastic	0696
1	Pipet, droping, plastic	0352
1	Nitroge (IV) oxide in air comparator	7689
1	Tubing	23609
1	Pipet	30410
1	Needle	27336-01

Table 4. LaMotte nitrogen (IV) oxide in air test kit code 7690

4.2.4 Nitrogen (IV) oxide lamotte sampling procedure

10mL of reagent #1 i.e. absorbing reagent was poured into the impinging tube, a gas bubbler impinger (0934). The impinging apparatus was connected to the intake of air sampling pump and the long tube was immersed in the absorbing solution. The special adaptor was attached to the intake of the pump to sample at $0.2Lm^{-1}$ while the sampling was done for 20 minutes when a measurable amount of nitrogen (IV) oxide was absorbed. At the end of the sampling period the contents of the impinging tube was poured into test tube (0822). The pipette (0352) was used to add a drop of reagent #2, the test tube capped and mixed after which the 0.05g spoon was used to add 0.05g of reagent #3. The test tube capped and the solution left for 10 minutes for colour development after which the test tube was placed into comparator (7689) and the sample colour matched to index of colour standards. The index number which gave the proper colour matched was recorded and the calibration chart used to convert the index read to concentration of nitrogen (IV) oxide in ppm.

	Comparator index number							
Time (min)	1	2	3	4	5	6	7	8
1	0.00	2.8	7.0	14.0	21.0	28.0	42.0	56.0
5	0.00	0.56	1.40	2.80	4.20	5.60	8.40	11.20
10	0.00	0.28	0.70	1.40	2.10	2.80	4.20	5.60
15	0.00	0.19	0.47	0.93	1.40	1.87	2.80	3.74
20	0.00	0.14	0.35	0.70	1.05	1.40	2.10	2.80

Table 5. Nitrogen (IV) oxide in air calibration chart**

4.3 Sulphur (IV) oxide

The absorbing solution used for trapping SO_2 was 0.3M H_2O_2 solution buffered at pH 5 ± 0.2.

The standard solution and calibration curve were prepared as follows:

0.1M H_2SO_4 was used as parent standard solution. All other lower concentrations were prepared from serial dilution of 0.1M H_2SO_4. 0.1M H_2SO_4 was standardized by titration against Na_2CO_3 using methyl orange as indicator. The conductivity measurement of each of the concentrations of H_2SO_4 (0.001 - 0.01M) obtained from serial dilution were taken, using Hanna Instrument EC 214 conductivity model. A graph of conductivity values in Siemens per centimeter (Scm^{-1}) against concentrations of H_2SO_4 in mol dm^{-3} was plotted. The data gave a straight line which passes through the origin with regression value of 0.9874. The calibration curve so obtained is shown in Figure 8. This was used as a working curve for the determination of SO_2 during the analysis of samples.

4.3.1 Sampling procedure

The procedure for sampling others remained except the flow rate that was increased to 2 $Lmin^{-1}$ for optimization purpose [Abdul Raheem et al., 2009[c]].

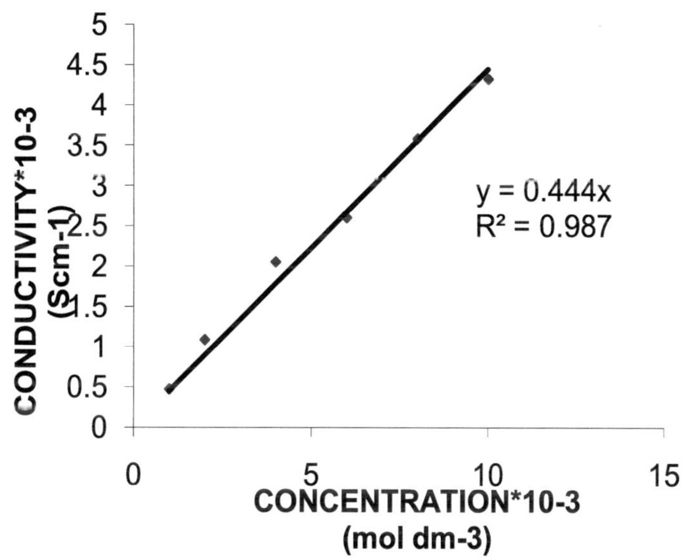

Fig. 8. Sulphur (iv) oxide calibration curve

4.3.2 Analysis

Conductivity measurements were undertaken using the Hanna Instrument Model E 214 conductivity meter.From the sample and reference solutions 20 cm^3 volume was measured respectively into a liquid sample holder test tube of Hanna model conductivity meter. The concentrations in mol dm^{-3} of H_2SO_4 formed from SO_2 of the samples were read out from the reference plot (Fig.8), the concentrations obtained in mol dm^{-3} were converted to parts per million or parts per billion or microgram per cubic meter (ppm or ppb or μgm^{-3}) as shown below using appropriate conversion factor. Equation of reaction for formation of H_2SO_4 from SO_2 is shown below:

$$SO_2 + H_2O_2 \rightarrow H_2SO_4 \quad (23)$$

4.3.3 Calculation [Stanley, 1975; Vowels and Connell, 1980]

$$ppm = \frac{moldm^{-3} \times mmSO_2 \times samplingvol.}{flowrate \times samplingduration} \quad (24)$$

$$\mu gm^{-3} = \frac{ppm \times molar\,mass}{24.45} \times 1000 \quad (25)$$

for 1 moldm^{-3} of SO_2 , ppm value will be

$$= \frac{1 \times 64 \times 30}{2 \times 60 \times 1000} \quad (26)$$

$$= 1.6 \times 10^{-2}\, ppm$$

in μgm^{-3} , the value becomes:

$$\frac{1.6 \times 10^{-2} \times 64 \times 1000}{24.45} = 41.88 \mu gm^{-3} \quad (27)$$

QUANTITY	CONTENTS	CODE
2 × 250 mL	Sulphur (IV) oxide absorbing solution	7804-K
15g	Sulphur (IV) oxide reagent #1	7693-E
30mL	Sodim hydroxide, 1.0 N	4004PS-G
60mL	Sulphur (IV) oxide passive bubbler indicator	7805-H
2	Pipets, 1.0mL, plastic	0354
2	Test tubes, 5 mL, plastic, w/caps	0230
2	Test tubes, Hester, w/caps	0204
1	Spoon, 0.25g	0695
1	Dispenser caps	0693
1	Sulphur (IV) oxide passive bubbler comparator	7746

Table 6. LaMotte sulphur (IV) oxide in air test kit code 7714

4.3.4 Sulphur (IV) oxide lamotte sampling procedure

10mL of Sulphur (IV) oxide absorbing solution was added to impinging tube and connected to the impinging apparatus as shown in Figure 5. The long tube was immersed into the absorbing solution. Sampling was done at 1.0 Lpm for 60 minutes or 90 minutes. The impinging apparatus was covered with foil to protect it from light. At the end of the sampling time the small test tube (0230) was filled to the line with the sample and 0.25g spoon was used to add a level measured of Sulphur (IV) oxide reagent #1. The test tube containing the mixture was capped and vigorously shaken to dissolve the powder. A 1 mL pipette was used to add 1mL sodium hydroxide, 1.0N, to the same small test tube, capped and inverted several times to mix. The other 1mL pipette was also used to add 2mL (2 measures) of Sulphur (IV) oxide passive bubbler indicator (7805) to a large test tube (0204). The contents of the small test tube were poured into the large test tube containing the indicator. Immediately the tube capped and inverted six times, holding the cap firmly in place with the index finger. After waiting for 15 minutes, the test tube was placed into the Sulphur (IV) oxide passive bubbler comparator (7746). The sample colour matched with the standard colour and the index number read and recorded from the comparator. The index number was converted to concentration in ppm using the calibration chart provided.

	Comparator index number							
Time (min)	1	2	3	4	5	6	7	8
10	0.00	0.19	0.29	0.38	0.48	0.57	0.67	0.76
30	0.00	0.06	0.10	0.13	0.16	0.19	0.22	0.25
60	0.00	0.03	0.05	0.06	0.08	0.10	0.11	0.13
90	0.00	0.02	0.03	0.04	0.05	0.06	0.07	0.08

** Values in pp

Table 7. Sulphur (IV) oxide in air calibration chart**

5. Quality assurance

The impinger was well rinsed with distilled water and properly wrapped with foil paper before each use. The tubing's and corks in the sampling train were checked before and during sampling, in case they had become slackened, however silicone grease was used to increase the pressure by making them air tight.

The absorbing reagents were always prepared freshly ahead of sampling for the solution to stabilize. They were stored in amber coloured bottles and refrigerated because of light interference. They were always allowed to thaw and assume the 25°C temperature before use. Lengthy contact with air by the absorbing reagent was avoided during both preparation and use to prevent absorption of the oxides. The absorbance of the reagent blank was deducted from that of the samples where the machine could not be adjusted to zero to avoid matrix error, especially with the conductivity meter.

For the nitrogen oxides determination, a gas bubbler impinger (fritted gas bubbler) was used instead of a general purpose impinger as absorption tube. The general purpose impinger has been reported to give low absorption efficiency with oxides of nitrogen [ICMA, 1972; Onianwa et al., 2001; Saltzman, 1954]. However the results were corrected and correlated with the fritted bubbler as well as standardized absorbing solution imported from LaMotte and Company, USA.

Greatest accuracy has been reported to be achieved by standardizing the sampling train with accurately known gas sample in a precision flow dilution system like a permeation tube [Dara, 2004]. Due to lack of the apparatus necessary for the standardization of the train, the actual collection efficiency is not known. However with the use of LaMotte sampling pump with inbuilt flow meter and standardized reagents, we recorded high collection efficiency at sites with increase concentrations of samples.

6. Results

This is already discussed extensively in Abdul Raheem, 2007 and Abdul Raheem et al., 2009[a,b,c]. Typical tables are shown to show the typical measurements concentration results and the meteorological data

Start of sampling	End of sampling	OX (ppb)	NOx (ppb)	SO_2 (ppb)	RELHUM (%)	WND ms^{-1}	DWND (°C)	AIRTEMP (°C)	Sun ExpWm^{-2}
6.30am	7.30am	29.08 ±11.73	1.47	7.83	78.17	27.60	144.60	22.70	-1.55
7.45am	8.45am	29.72 ±10.5	3.44	6.54	71.67	36.30	156.40	23.20	0.51
9.00am	10.0am	29.71 ±5.57	0.43	4.17	57.30	44.60	156.50	27.90	8.63
10.15am	11.15am	33.11 ±5.51	1.67	4.42	53.30	42.00	160.50	29.80	12.61
11.30am	12.30pm	46.69 ±7.49	1.73	6.27	42.00	42.30	153.20	31.50	15.36
12.45pm	1.45pm	69.94 ±15.45	1.04	7.36	38.67	43.40	154.00	32.80	16.09
2.00pm	3.00pm	35.55 ±11.21	2.46	8.84	35.50	41.60	160.00	34.30	13.39
3.15pm	4.15pm	21.44 ±6.31	2.46	7.62	37.17	39.40	167.90	33.80	10.16
4.30pm	5.30pm	17.62 ±3.13	2.69	9.52	39.00	39.30	178.00	33.00	5.66
5.45pm	6.45pm	11.56 ±2.19	2.91	9.11	42.67	37.60	176.70	31.30	0.86

Table 8. Dry season environmental data for Ilorin

Start of sampling	End of sampling	OX (ppb)	NOx (ppb)	SO_2 (ppb)	RELHUM (%)	WND ms^{-1}	DWND (°C)	AIRTEMP (°C)	Sun Exp Wm^{-2}
6.30am	7.30am	14.26	12.40	10.72	90.58	3.07	107.02	25.99	-1.65
7.45am	8.45am	22.92	5.89	7.20	87.58	4.67	156.38	26.38	2.80
9.00am	10.0am	28.95	5.39	11.15	73.75	7.09	189.02	29.05	8.90
10.15am	10.0am	46.86	5.66	14.82	67.92	7.35	182.63	30.25	11.00
11.30am	12.30pm	43.21	6.41	10.51	63.50	8.76	170.67	31.30	12.06
12.45pm	1.45pm	85.31	5.68	12.74	60.33	10.11	159.55	32.00	17.30
2.00pm	3.00pm	73.77	6.45	16.62	60.08	10.36	155.00	31.98	15.10
3.15pm	4.15pm	26.06	6.84	15.47	62.67	10.94	163.79	31.37	13.20
4.30pm	5.30pm	12.23	5.72	16.48	67.00	10.21	165.22	30.38	10.70
5.45pm	6.45pm	8.58	6.90	19.21	72.75	8.99	166.64	29.20	3.30

Table 9. Dry season environmental data for Lagos

This is showing typical results of statistical modeled analysis of Ilorin and Lagos during dry season MLR with backward selection in stepwise mode (without intercept) results in the following equation:

$$OX_{ILO} = 6.092 \times SO_2 + 0.657 \times RHUM - 2.653 \times ATEMP + 4.385 \times SUNEXP \quad (28)$$

Where R = 0.981, F (4, 6) = 38.389, $p < 0.000$

This shows that only four of the variables are found to be significant for retention in the model.

MLR using backward selection in stepwise mode (without intercept) results in the following equation:

$$OX_{Lag} = 1.679 \times ATEMP + 5.622 \times SUNEXP - 8.079 \times WND \quad (29)$$

where, R = 0.961, F (3, 7) = 27.874, $p < 0.000$

MLR shows that only three of the variables are significant for retention in the model.

A table Comparing the ozone measured concentration with calculated results from MLR model equations

	ILORIN		LAGOS	
	RAIN	DRY	RAIN	DRY
MEASURED	21.86 ± 2.47	32.44 ± 5.13	9.87 ± 0.99	36.22 ± 5.76
MODELED	16.12 ± 1.86	44.32 ± 4.25	9.89 ± 0.82	36.29 ± 3.87

Table 10. MLR equation modeled results for ozone compared with monitored results for the two cities of interest during rainy and dry seasons (ppb)

7. General conclusion

The direction and spatial extent of transport and the relative contribution of transported ozone and precursors to individual downwind areas are highly variable. A number of factors influence site to site differences in ozone concentrations, including sources of precursor's emissions and meteorological conditions.

Data analysis also reveals that NO_x and SO_2 as well as volatile organic compounds contribute to ozone formation and this is in accordance with other researchers [Winer *et al*, 1974; Canada - US, 1999; chou *et al*, 2006]. The relative effectiveness of reductions of these three precursors can vary with location and atmospheric condition. Overall the concentrations of ozone could be said to be influenced globally by background concentrations, locally generated concentrations and transported concentrations.

On the whole the chemometric multivariate analysis results confirmed our experimental results and unfold the fact that meteorological influence plays a major role in the atmospheric chemistry of ozone.

Finally, these results and analysis suggested that ozone acting in concert with other pollutants need to be recognized as important health and ecosystem related air quality concern in Nigeria. Based on increasing evidence on regional transport of ozone all over the world, there is need for recognition that ground - level ozone would be an appropriate issue to be considered by the Nigerian government. In particular, a proactive measure has to be formulated towards reducing NO_x and SO_2 and by consequence O_3 in Nigeria.

8. References

Abdul Raheem, A.M.O.; Adekola, F.A.; Obioh, I.B. (2009) *Bull. Chem. Soc. Ethiop.*, 23 (3), 383 - 390[a]

Abdul Raheem, A.M.O.; Adekola, F.A.; Obioh, I.B. (2009) *Environ. Model. Assess.* 14:487-509[b]

Abdul Raheem, A. M. O; Adekola, F. A; Obioh, I. B. (2009) *SCIENCE FOCUS*, 14 (2)166 - 185[c]

Abdul Raheem, A.M.O. (2007) *Ph.D. Thesis, University of Ilorin*, Nigeria

ACGIH (1991) American Conference of Governmental Industrial Hygienists *Documentation of Threshold Limit Values and Biological Exposure Indices*, Vol. 2 pp 786-788. 6th ed., ACGIH Cincinnati

Alvarez, E., Pablo, F., Thomas, C., & Rivas, S. (2000) *International Journal of Biometeorology*, 44, 44–51.

Bakken, G. A., Long, D. R., & Kalvis, J. H. (1997) Examination criteria for local model principal component regression. *Applied Spectroscopy* 51, 1814–1822.

Bloomfield, P., Royle, J. A., Steinberg, L. J., & Yang, Q. (1996) *Atmospheric Environment* 30, 3067

Chou, C. C.-K., Liu, S. C., Lin, C.-Y., Shiu, C.-J., & Chang, K.-H. (2006) *Atmospheric Environment* 40, 3898–3908.

Cox, W. M., & Chu, S. H. (1992) Atmospheric Environment, 27B, 425.

Dara, S.S. (2004) *A Textbook of Environmental Chemistry and Pollution Control*, S. Chand and Company Ltd. New Delhi 110055

ECE (1984) *Effects of acidifying depositions and related pollutants on forest ecosystems*, Executive Body for the Convention on Long-Range Transboundary *Air Pollution*, Third Session, Geneva; 5-9 March.

Einax, J. W., Zwanziger, H. W., & Geiss, S. (1997) Chemometrics in environmental analysis. Weinheim: Wiley

Jackson, J. E. (1991) A user's guide to principal components New York: Wiley

Jaffe, D. A. T., Covert, D., Kotchenruther, R., Trost, B., Danielson, J., Simpson, W., (1999) *Geophysical Research Letters* 26, 711–714.

Klaus, D., Poth, A., & Voss, M. (2001) *Atmosfera* 14, 171–188.

Lengyel, A., Heberger, K., Paksy, L., Banhidi, O., & Rajko, R. (2004) *Chemosphere* 57, 889–896.

Menezes, K. A. & Shively, T. S. (2001) *Environmental Science & Technology* 35, 2554– 2561.

Norman, C. (1987) Analyzing multivariate data. San Diego, CA: Harcourt Brace Jovanovich

Onianwa, P.C.; Fakayode, S.O.; Agboola, B.O. (2001) *Bull. Chem. Soc. Ethiop.* 15, 71

Otto, M. (1999) Chemometrics statistics and computer application in analytical chemistry. Weinheim: Wiley-VCH

Pissimanis, D. K., Notaridou, V. A., Kaltsounidis, N. A., & Viglas, P. S. (2000) Theoretical and Applied Climatology 65, 49–62

Schutt, P.; Cowling, E.B. *Plant Disease* 1985, 69, 548.

Shively, T. S., & Sager, T. W. (1999) Environmental Science & Rechnology 33, 3873

Timm, N. H. (1985) Applied multivariate analysis. Englewood Cliffs, NJ: Prentice-Hill

Vautard, R., Builtjes, P. H. J., Thunis, P., Cuvelier, C., Bedogni, M., Bessagnet, B., (2007) *Atmospheric Environment* 41, 173–188

Vogt, N. B. (1989) Chemometrics and Intelligent Laboratory Systems 7, 119–130

Watson, J. G. J. (1984) Air Pollution Control Association 34, 619–623

World Health Organization (WHO) (1981) Sulfur dioxide, *Environmental Health Criteria*, WHO: Geneva

Spectroscopy Analysis of Corrosion in the Electronic Industry Influenced by Santa Ana Winds in Marine Environments of Mexico

Gustavo Lopez[1], Benjamin Valdez[2] and Michael Schorr[2]
[1]*CETYS Universidad, Departamento de Cibernética-Electrónica,*
[2]*Instituto de Ingeniería, Departamento de Materiales, Minerales y Corrosión,*
México

1. Introduction

Climate change in some regions of the world is due to the effect of variations meteorological phenomenon, such as El Niño southern Oscillation (ENOS), which occasioning rainfalls in winter and even flooding, cold fronts and tropical cyclones in Baja California. Santa Ana winds (SAW) are influenced by ENOS, originated in the Santa Ana Canyon in the Mojave desert (Traviña et al, 2002), which cause rapidly changes in the climate conditions in the south west of California, USA y northwest of Baja California, Mexico. SAW are developed when the desert is cold, and are presented most commonly during autumn and spring seasons. This originates fast temperature rises and relative humidity (RH) drops, causing damage in the vegetation of these zones and changes in the meteorological conditions affecting the environments in indoor of industrial plants. Due to drastic changes in temperature and humidity in indoors of companies by SAW, an effect of strong transitions of these climatic factors occur which in combination with air pollutants as sulphurs and chlorides generate deterioration of copper metals of electronic machines and equipments. Analysis of raw materials used in electronic devices and electrical failures of electronic components were carried out in 15 companies in the coast of Baja California, Mexico. This study presents an analysis of winter of 2009, early spring of 2010 and late autumn of 2010. Humidity and temperature ranges of these winds can change rapidly from dry to wet, cold to warm and reverted in some hours. These several variations influence the corrosion rate of copper electrical connectors and connections of electronic equipments near the coast, installed in industrial parks of the northwest of Mexico in marine environments. In this region are located two important cities: Tijuana in the border with San Diego, California where was evaluated this phenomenon in ten companies and in Ensenada at 100 km. near the USA-Mexico border, with five industrial plants analyzed. The amount of defective devices increased in 25% and failures in electronic machines and equipments in 28% in the period of SAW occurred, compared with other seasons. In addition, 100% of raw materials located in warehouses of industrial companies, 33% were unusable. Auger electron spectroscopy (AES) technique was applied to determine the corrosion products in electrical connections and connectors of industrial electronic devices, equipments and machines.

1.1 Electronics industry

The electronics industry has grown over the past fifty years, contributing to the economy. Particularly in the Baja California State located in the northwest of Mexico, these companies have prospered in the industrial parks of Tijuana and Ensenada, on the Pacific Ocean considered as a marine region. In this environment, during winter and summer, the climate main factors at indoor conditions: humidity and temperature, affect electronic devices and industrial machines (Lopez et al, 2007). The change of climate in indoor of electronic plants is due to the variation of humidity, temperature, solar radiation, concentration of air pollutants such as CO, SO_2, H_2S, NO_X, O_3 and solid particles $Pm_{2.5}$ and Pm_{10}. The air pollutants and solid particles are monitored by Environmental Monitoring Stations (EMS) in Tijuana. While, SO_X and Cl^- were determined in Ensenada by Sulfatation Plates Technique (SPT) and Wet Candle Method (WCM), and metallic probes and in both cities were analyzed by AES. Electronic equipments installed inside the plants are fitted with copper components since they exhibit electrical and thermal conductivity. With the deterioration of copper, the equipment functionality decreases, generating failures and economic losses. This equipment is exposed to a wide range of indoor aggressive environments causing corrosion damage.

1.2 Atmospheric pollution

Gaseous and solid airborne pollutants comprises small dust particles hydrogen sulfide (H_2S), sulfur dioxide (SO_2), chloride ions (Cl^-) and nitrogen oxides (NO_X) which are capable of penetrating into indoor areas of the industrial plants through inlets of air filtering systems. These pollutants generate aggressive indoor environments in combination with humidity and temperature, producing corrosion damage of metallic components of electronic equipment. In spite of the existing corrosion prevention and protection system as well as application of electronic industry standards, corrosion control is not easy in specific climatic regions, especially in arid and marine regions (Dillon, 2000). Arid zones are typical, with RH ranging from 30% to 90% in summer with temperatures higher than 40°C and winter temperatures lower than 5°C. In marine environments, the RH is around 20% to 80% and temperatures in summer arises 35°C and diminishing to 3°C in winter. These factors, promoted by anthropogenic and natural activities facilitate the corrosion of the copper components (Valdez et al, 2006). The corrosion of copper specimens was evaluated by the gravimetric method; it was correlated with the minimum, average and maximum RH and outdoor temperature values in different seasonal periods, which affect the indoor climate plants. The efficiency of the electrical behavior of copper components in electronic devices is a function of the amount of humidity and pollutants present in the indoor environment (Veleva et al, 2008).

1.3 SAW and their effect on climate factors

A weather phenomenon in the region of California, USA and Baja California, Mexico, is SAW, which appears in finals of autumn and early of winter and spring (Traviña et al, 2002). This leads sometimes drastic changes in humidity and outdoor temperature affecting the indoor environments of industrial plants, originating in the manufacturing processes failures in electronic devices and equipments, causing low yielding. Moisture is the most important climatic factor that has a negative effect of this phenomenon, decreasing in certain periods of this meteorological phenomenon from values of 80% to 30%. SAW pass through the mountain and sometimes reach the hurricane force. The combination of wind, heat, and dryness generates dry vegetation and thus is an explosive fuel for fires acquaintances and often devastates the region. Though winds are often of a destructive nature, may have some positive results too. They can make the cold water rises from the ocean bottom to top, dragging many nutrients that ultimately benefit the fishery. As winds blow over the ocean, the temperatures of the sea surface fall around 4 ° C (7 °F), indicating an upwelling of deep ocean water. The evaporation process originated by SAW generates increases of temperatures by about in 5 ° C. These variations modify the physicochemical properties of metallic materials and occur in some times a partial deterioration and in a few times after the beginning of SAW originates the corrosion in metallic materials as copper. This concerned to companies for the economic losses of defective electronic products and promotes rapidly the corrosion process on connectors and electrical connections and thus causing electrical failures, malfunctioning in electronic equipment and in some cases stops the manufacturing line. This causes warm climates, causing deterioration in some metallic materials used in the electronic connectors and connections such as copper. This decrease the quality of the articles manufactured in this region and generates economic losses (Lopez et al, 2007). When variations of humidity are very drastic, occurs several variations of temperature and therefore the electronic equipments and devices suffer changes in their functionality. To measure humidity ranges, is necessary install sensors in indoor of industrial plants be in contact with environments.. This study in this region is important by the influence of pressure difference of air mass between the mountains of California, United States and marine areas of northwestern of Mexico country, where companies are located in these the cities: Ensenada (300,000 people) and Tijuana (2,000, 000 people) (CONAPO, 2010). An analysis of wind direction and speed was made in the seasons studied to correlate this climate factors with the electrical failures of electronic equipments (Table 1).

Month	Wind Direction (°)		Wind Speed, m / s	
	Ensenada	Tijuana	Ensenada	Tijuana
January	66	72	14	11
February	69	70	16	13
March	67	69	14	12
April	61	63	18	14
May	223	228	10	7
June	228	234	11	9
July	192	198	14	10
August	195	201	13	11
September	206	211	10	8
October	204	207	11	8
November	63	69	8	6
December	67	74	9	6

Table 1. Analysis of wind in Ensenada and Tijuana (2010)

Wind direction from January to April and November to December was from southwest to northeast (60° to 70°), and in May to October, winds were reverted to the same direction (180° to 240°) and wind speed was at velocities less of 20 m/s. In the months that SAW is presented, RH decreased from 95% to 40% and temperature increased from 35 °C to 20 °C very fast in some hours in indoor of industrial plants.

1.4 Climate factors vs. electrical failures

Electronic equipment and machines of industrial plants consist of printed circuit boards, also called e-cards. The lifetime of these electronic components is according to the type of material, operation and exposition to environment. In some occasions, the durability decreases by the exposition to air pollutants and climatic factors, mainly drastic variations of humidity and temperature (Veleva et al, 2008, Lopez B.G. et al., 2010). Engineering design, manufacturing methods wrongs and inadequate mechanical movements generate improper installations. E-cards are constituted of a lot of electronic devices and are installed according to their operation, with sections separated with insulators to avoid electrical connections between undesirable zones (analog or digital signals), power (high and low), speed (high and low) and types of printed boards. There are a variety of electronic components, which are classified according to wave (sine, triangular, square or linear). In this study, an analysis of climate factors and electrical failures of electronic equipments installed in indoor of industrial plants of these marine zones was made (Figure 1).

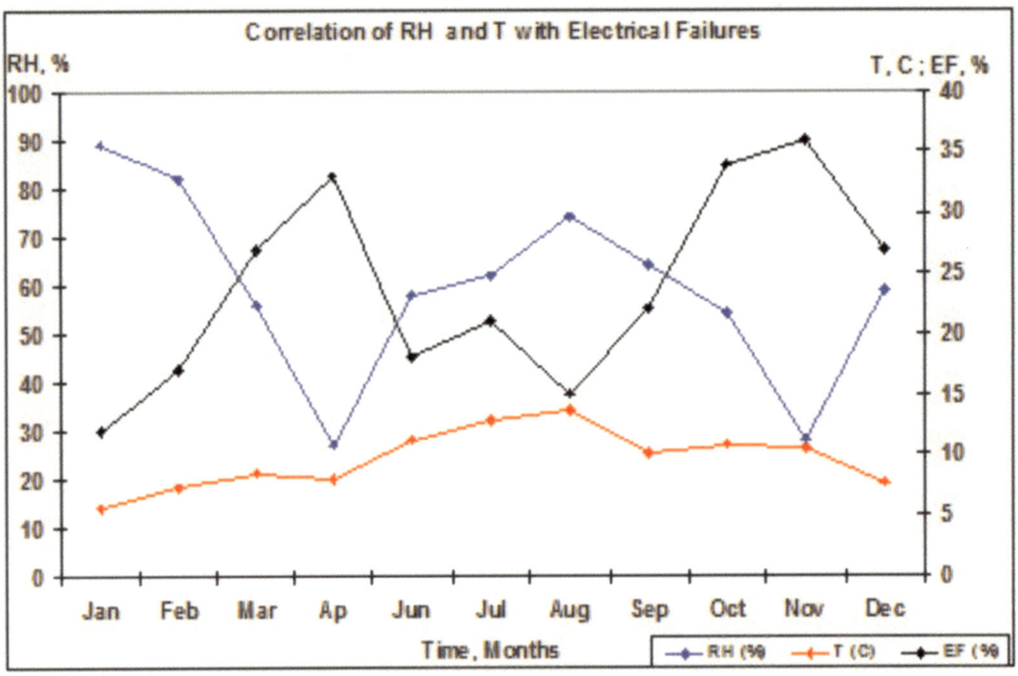

Fig. 1. Correlation of climate factors with electrical failures in industrial plants (2010).

In the absence of SAW, values of RH were in the range of 80% to 90% and temperatures variety from around 15 °C to 30 °C and electrical failures by an average of 20% at month.

When SAW were presented, RH diminishes to around of 30% and temperatures increases by 5 °C and occurring electrical failures that are increased around 50%.

2. Materials and methods

Copper is used in electronic devices and equipments for the manufacturing of electrical connections and connectors, t for its good electrical and thermal conductivity. Nevertheless, is very susceptible to corrosion at the indoor polluted conditions.

2.1 Climate factors

Moisture content for interior environments modifies the physicochemical properties of metallic materials and increases strongly their micro and nano deterioration by the variation of air moist to dry, causing warm climates very quickly (Zlatev et al, 2009, Cole, 2004). Climate is composed of several parameters, where the relative humidity and temperature are the most important in the damage to these materials (ASHRAE, 1999), expressed with their measurements units and instruments (Table 2). Scientists that analyze deterioration of materials and climatic conditions in indoor of industrial plants, consider that a bad operation of industrial equipments is due to the presence of drastic changes in humidity and temperature in certain times of the year (Lopez, 2008), as is expressed in ISO 9223 (ISO9223, 1992). Experts in industrial productivity and quality manufacturing consider that in these region defective products and electrical failures in electronic equipment of industrial plants are presented and create a tense in managers in the companies because causes great economic losses.

Factors	Measuring Instrument	Unit
Humidity	Hygrometer	%
Temperature	Thermometer	◦ C
Atmospheric pressure	Barometer	mmHg
Solar radiation	Pyranometer	W / m2
Pluvial Precipitation	Rain gauge	mm
Wind direction	Wind vane	◦ Grade
Wind speed	Anemometer	m/s

Table 2. Climatic factors and their measurement

2.2 Corrosion testing

Copper specimens for corrosion testing were exposed in at indoor conditions of electronics plants for a period of two years in Mexicali and in Ensenada. The specimens corrosion was evaluated according to the ASTM standards G 1, G 4, G 31 (ASTM, 2000), and correlated with the minimum, average and maximum outdoor RH, TOW and temperature parameters (ISO 11844-1, 2006). The electronic plants in Tijuana where they are located in industrial parks while in Ensenada, plants are located at distances ranged between 1 km to 10 from the sea shore. Rectangular specimens with dimensions of 2.5 cm. x 1 cm. x 0.5 cm. were cleaned by immersion in an isopropyl alcohol ultrasound bath for 15 minutes. Immediately after cleaning the copper specimens were placed in sealed plastic bags, ready to be installed in the test indoor sites. After each exposure period the specimens were removed, cleaned and weighed to obtain the weight loss and to calculate the corrosion rates (CR) (ISO 11844, 2005). The corrosion products morphology was examined with an optical microscope. The most active gases are hydrogen sulfide (H_2S) and SO_X. H_2S is a corrosive and toxic pollutant which originates from municipal sewage (Moncmanova, 2007; Lopez B. et al, 2010).

2.3 Numerical analysis

A mathematical correlation was made in MatLab software (Duncan et al, 2005) to determine the corrosivity levels (CL) in indoors of electronics industry in Mexicali an Ensenada, With this analysis find out the deterioration grade (DG) of metallic probes of copper, correlating the climate factors (humidity and temperature), airs pollutants (CO, NO_2, O_3 and SO_2), from outdoor sources that penetrate by inlets or air-conditioning systems, to indoors of industrial plants and corrosion rate (CR).

2.4 Characterization of atmospheric corrosion in industrial plants

Corrosion in indoor atmospheres is influenced by external climatic conditions generating the time of wetness (TOW). TOW is an important factor in the generation of corrosion in electrical connectors and connections of electronic devices, was obtained with variations of RH and temperature, showing high rates in some periods of winter and summer by the SAW phenomena. This is an important factor to determine the type of corrosion that occurs in metals and the levels which increases the corrosion rate (CR). In this process, cells formed by corrosion process, originate galvanic corrosion, for example in the binding of platinum and zinc with 0.2 volts, which are capable of cause an electrical failure in electronic devices and equipments. The origin of corrosion in indoors of industrial plants is due to climatic factors and the concentration of air pollutants. In winter, corrosion is generated uniformly by wetting film formed on metals, and in summer in the materials analyzed appears small tarnishings, as a result of some air pollutants and drops adhered to metals surface. The TOW obtained of variations of RH and temperatures with values over of 80% and 0 ° C, that is characteristic of three cities evaluated, is an important factor in the origination if corrosion in metals. All regions analyzed, showed corrosivity levels (minimum and maximum), that are a cause of generation of aggressive environments, and the corrosion process. The period of copper exposure, appears small tarnishing in some areas of metallic probes, which caused pitting and uniform corrosion (Lopez , 2008).

2.5 Spectroscopy examination techniques

The examination technique with the Auger Electron Spectroscopy (AES) was used to determine the air pollutants that reacted with the copper surface of metallic probes and the morphology with the Auger map (Yves Van, 2008). Electrons flows through the instrument to generate low energy showed the topography of metals tested. Once it determines the chemical composition of elements and compounds in the specimens, and observed the air pollutants deposited on metals and know that they affect in the generation of the corrosion process. With this technique we knew in detail, quickly and with precision, the structural form and location of corrosion, which will determine the type of corrosion and know the techniques protection for the metals analyzed (Asami et al, 1997, Briggs, 1990)). The specimens were exposed in indoor of industrial plants with values of relative humidity and temperature over 90% and 20 °C in winter and 80% and 15 °C in summer, and removed for one, three and six months and analyzed with EDX and AES spectroscopy to identify the corrosion products formed on the copper surface. Metallic probes were exposed in periods of 1, 3 and 6 months showed tarnish in some regions, but from the seventh month to the final period of study, in the specimens appear corrosion on the full surface of copper, and corrosion rate decreases by the formation of a copper oxide film, serving as a protective layer. The analysis obtained by EDX and Auger S2 models was made with Ranger and ESCA / SAM 560 of Perkin Elmer. The Auger spectra showed the chemical composition and use the sputtering method to clean the copper surface, bombarding with Ar+ ion beam with energy of 5keV and current density of 0.3 uA / cm^3, to remove the CO2 pollution of the environment.

3. Results

The rapid transitions of temperature and humidity in indoor of companies caused by SAW and generate aggressive environments in combination with air pollutants generated in indoor of industrial plants and stays longer periods of time suspended in the air and causing more damaging in electronic equipments.

3.1 Deterioration of metals used in industrial plants

Ranges of RH and temperature were higher than 70% and 35°C during the year in Tijuana and with a minimum of 20% and 30°C in the periods of heat winds (early winter and spring) and maximum of 80% and 10°C in the rest of the year in Ensenada area. Levels of humidity and temperature bigger 70% and 30 °C accelerated the rate of corrosion. In summer the corrosion rate was higher after one year in Tijuana and also was in Ensenada. For temperatures in the range from 25 °C to 35 °C, and RH level of 30% to 70%, the corrosion rate was very high. Furthermore, in winter, at temperatures around 15 °C to 25 °C and RH levels from 35% to 75%, water condensates on the metal surface and the copper corrosion rates increases very fast. Variations of RH in the range from 30% to 80% and temperatures from 0 °C to 35 °C, and concentrations of air pollutants such as sulfides and chlorides in this marine environments, exceeding the permitted levels for air quality standards, are also an important condition that favors corrosion. Corrosion processes can be accelerated in polluted atmospheres in the presence of variations of humidity and temperature. The origin of corrosion in metallic materials such as copper, used inside the electronics industry is due to the action of climatic factors and effect of air pollutants. In the autumn and winter, corrosion is generated by a film formed uniformly on metals, and in spring in these materials are pitting as a result of some small drops that adhere to certain areas of the metal surface (Lopez et al, 2010). This is partly of the variations of humidity and temperatures that is typical of this city during certain seasons of the year. Is necessary monitoring and control principally the relative humidity and hold the adequate range to permit the good functionality of the equipments and products to avoid defective products and electrical and mechanical failures of electronic equipments during the presence of SAW. The values of relative humidity and temperature were lower 40% and 20 °C during the year. At beginning of spring season, the corrosion rate increases in the range 15 °C to 30 °C, with relative humidity level of 25% to 40%. Furthermore, in autumn and winter the several changes of RH and temperature, water condensates on the metal surface and the copper corrosion rates increases. At temperatures of 17 °C to 25 °C, with the variations drastically of RH levels, corrosion rate was high too (Table 2). The emissions of vehicles are an important factor source of air pollutants, are cars that is important in the marine zone and emissions of thermo electrical operations where the electricity is generated to this. The air pollutants affect on the metals and the corrosion rate depend on the pollutant concentration in air and their corrosion resistance in the region. The Cu_2O layer can be gradually dissolved in the acidic electrolyte formed in the presence of SO_2 and Cl^-. A computer based model of atmospheric corrosion has been used to simulate copper exposed to room temperature to 200 ppb SO_2, in combination with either NO_2 or O_3 at different concentrations.

Climate factors	Sulphur oxide (SO_2)				Nitrogen oxides (NO_X)				Ozone (O_3)			
	RH[a]	T[b]	C[c]	CR[d]	RH[a]	T[b]	C[c]	CR[d]	RH[a]	T[b]	C[c]	CR[d]
Spring Max Min	74.5 22.7	22.3 10.2	0.21 0.18	225 17	79.3 29.6	29.5 11.7	0.42 0.28	208 109	69.7 21.5	24.3 12.5	0.29 0.12	193 145
Autumn Max Min	86.4 26.4	28.6 19.4	0.27 0.16	289 174	73.9 25.4	31.7 12.7	0.44 0.19	265 168	74.9 18.7	29.5 14.8	0.52 0.18	268 203
Winter Max Min	69.2 19.3	22.6 13.3	0.48 0.32	324 152	85.1 28.6	25.7 14.8	0.56 0.29	231 188	79.9 26.4	28.9 12.8	0.67 0.23	273 165

[a] RH. Relative Humidity (%) [b] T. Temperature (°C)
[c] C. Concentration Level of Air Pollutant (ppm)
[d] CR- Corrosion rate (mg/m^2.year).; Source. Environmental Monitoring Stations.

Table 3. Effect of RH, temperature and air pollutants on the corrosion rate of copper in Tijuana.

3.2 Corrosivity levels analysis

Corrosivity levels (CL) represents the degradation grade (DG) of materials in according to the correlation of relative humidity, temperature and air pollutants concentration mentioned above. CR was calculated from the atmospheric pollution in indoor of electronics industry, indicating which air pollutants with major grade of damage to copper was dioxide sulphide (SO_2) and the ion chlorides (Cl^-) in both cities.

In both cities, RH was correlated with the major CR was 30% to 50% with temperatures of 25°C to 40°C. In summer CR was different than in winter, and in both environments (Figures 2 and 3). The electronics equipment corrodes at high humidity levels. Air pollutants such as CO, NO_X and sulfides penetrate through defects of the air conditioning systems. Corrosion phenomena affect connections of electronic equipment and other electronics components protected with plastic or metallic materials (Sankara et al, 2007). Atmospheric corrosion is an electrochemical phenomenon that occurs in the wet film formed on metal surfaces by climatic factors in Tijuana was presented pitting corrosion and in Ensenada was presented

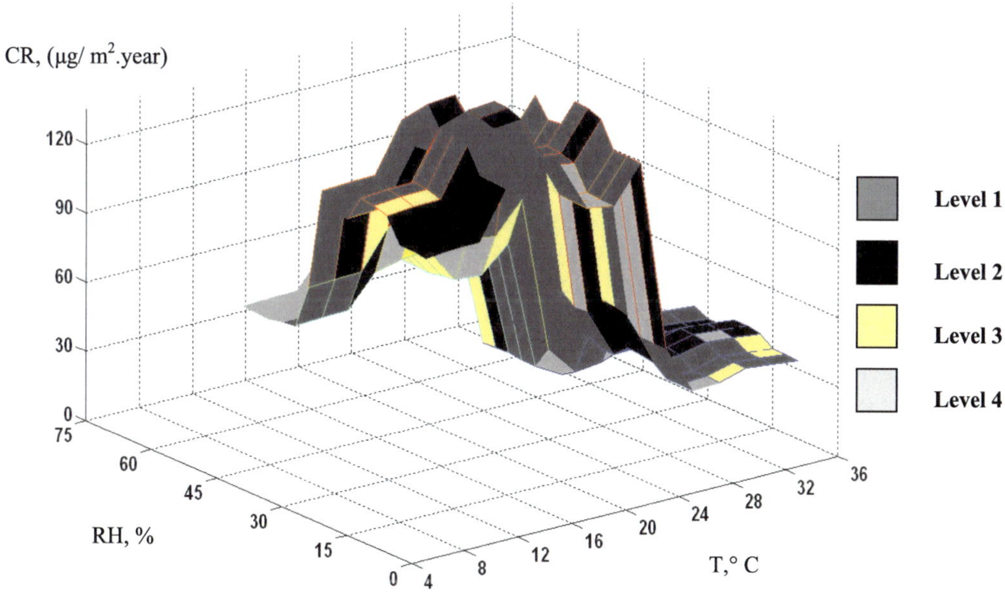

Fig. 2. CL of copper after one year of exposition in Ensenada (2010).

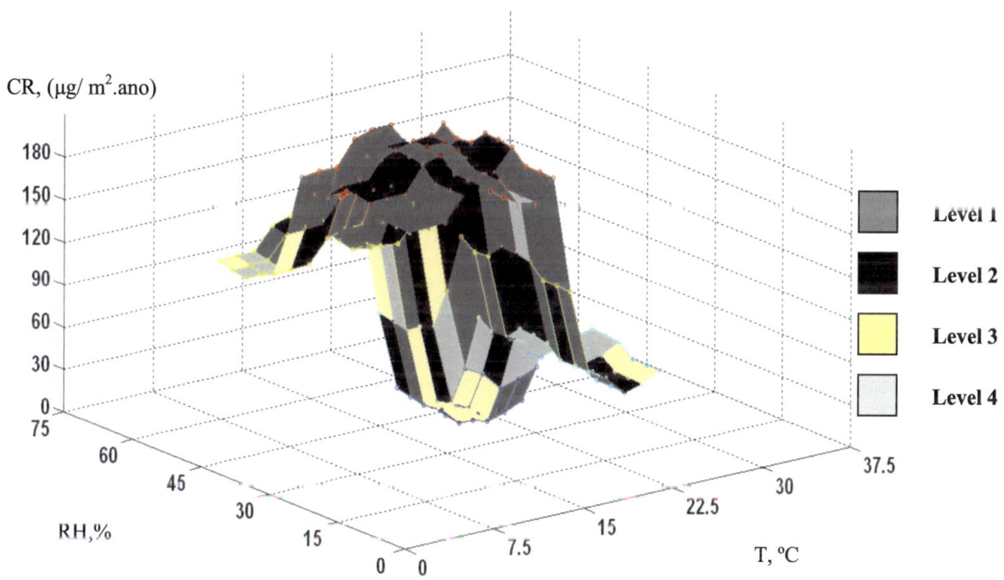

Fig. 3. CL of copper alter one year of exposition in Tijuana (2010).

the uniform corrosion. This behavior of corrosion was showed in the first month of the period of evaluation. Figure 2 shows the CL of indoors of electronics industry in Ensenada, indicating the level1, the major aggressive environment and levels 4 the high aggressiveness grade (AG) which generate high DG of this type of materials. Some sections of the graph, represents the different grades of AG, with high areas of level 1 and 2 and levels 3 and 4 exists in less percentage. This can generate failures in the electronic devices and equipments in indoor of industrial plants. RH and temperature ranges were from 30% to 75% and 20 ° C to 35° C with CR from 30 µg/m^2.year to 100 µg/m^2.year. The same was occurred in Tijuana city that is showed in figure 3. Evaluation of CL in Tijuana indicate that same range of levels presented in Ensenada, with RH and temperatures from 40% a 75% and 20 ° C to 35 ° C, with CR from 10 to 160 µg/m^2.year. Microcircuits, connectors and electrical contacts used in the electronics industry, are susceptible to atmospheric corrosion, which occurs in indoor plants in the industrial plants in Tijuana and Ensenada. Corrosion rates obtained in semi-arid testing points (Tijuana) is higher than those from the marine environments (Ensenada), since in the arid zone the principal air pollutant is H_2S and in the marine zones, predominate the Cl^- ion as the major corrosive. A comparative analysis, after six months of exposure of copper specimens shows higher levels of deterioration by corrosion in Tijuana regarding to the Ensenada results.

3.3 AES examination

AES analyses were carried out to determine the corrosion products formed in the copper surface. Figure 1a show scanning electron micrograph (SEM) images of areas selected for AES analysis covered by the principal corrosion products which are rich in chlorides and sulphides in metallic specimens evaluated in both cities. The Auger map process was performed to analyze punctual zones, indicating the presence of Cl- and S as the main corrosive ions present in the copper corrosion products. The Auger spectra of Cu specimens was generated using a 5keV electron beam (Clark et al ,2006), which shows an analysis the chemical composition of the thin films formed in the Cu surface in Ensenada (Fig 2) and in Tijuana (Fig 3). The AES spectra of copper specimens installed in industrial plants in both cities show the surface analysis of three points evaluated in different zones of the metallic probes. The peaks of Cu appear between 905 and 915 eV, finding the chlorides and sulphides. In figure 2, the spectra reveals the presence of carbon and oxygen, chlorides and sulphides, with variable concentration in the chemical composition in the three regions analyzed, where the principal pollutant was Cl- ion. Figure 3 corresponds to the specimens installed in companies in Tijuana city. In the regions of copper surface analyzed were observed different concentrations of sulfur, carbon and oxygen, while the main air pollutant detected was H_2S. The atomic concentration (%) of the chemical elements in each spectrum was organized in Table 4. To draw this figures a program: sigma plot was applied. The spatial resolution of this technique is around 100 nm and a 1 nm depth resolution (Swart, 2010).
The metallic samples of 1, 3 and 6 months to the metals analyzed show localized corrosion with small spots and from the seventh month to the one year the spots were observed larger, being more concentrated corroded areas and generated uniform corrosion. Auger spectra show an analysis of the chemical composition and a microphotograph on the metals surface (Figure 4) and with the sputtering [14, 15] was observed the corrosion and passivity process that generate the copper oxide film to reduce corrosion. Air pollutants added to copper surface form thin layers in according to the adhesion grade, in some with more chloride ions (light color) and other with less concentration of sulphur (dark color) for the marine environment in Ensenada (Figure 4a) and elements with carbon oxide of the environment by the sensibility of the Auger equipments was showed in the spectra (Figure 4b). In the analysis of Auger spectra in Tijuana that is an arid environment, the major air pollutant that reacts with copper surface was the sulphur, represented by the Auger map (Figure 5a) with

more dark sections without chloride ions, compared with the evaluation in Ensenada. Auger spectra indicate the air pollutants combined with CO_2 of the environment (Figure 5b). Table 4 represents the atomic concentration of metallic probes with the percentages of air pollutants and carbon and oxygen (Sankara et al, 2007, Clark, 2006).

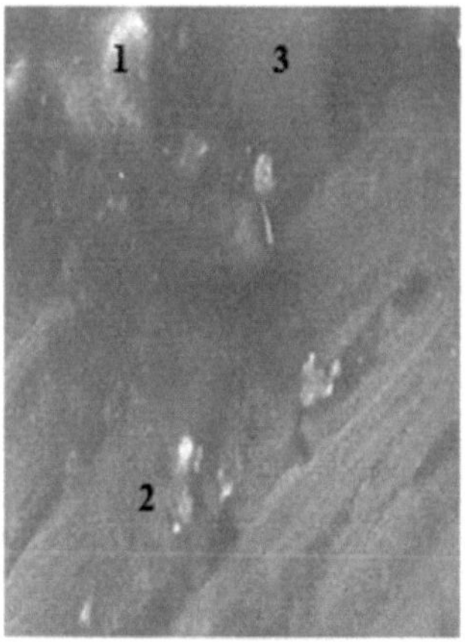

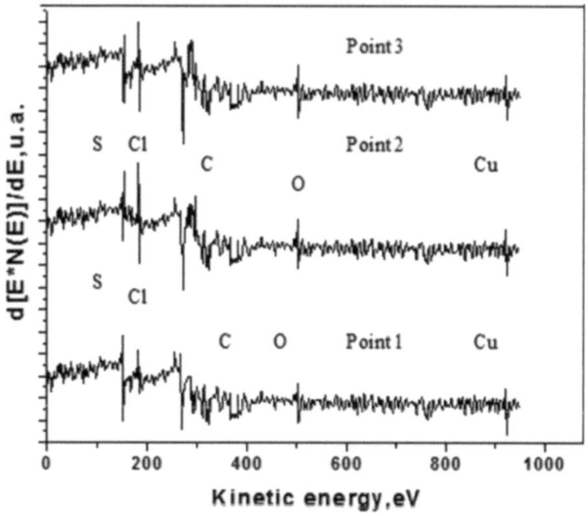

(a) (b)

Fig. 4. Analysis of corrosion products of copper at one year of exposition: (a) Auger map and (b) AES in Ensenada (2010).

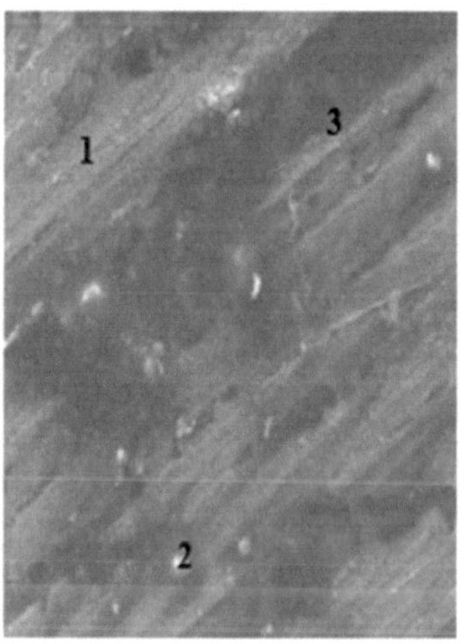

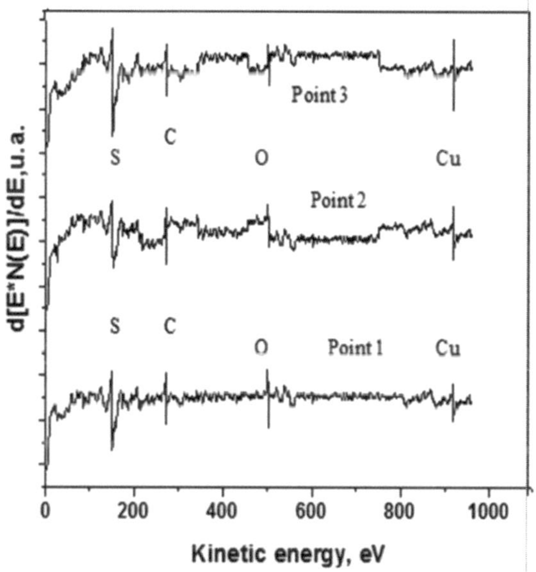

(a) (b)

Fig. 5. Analysis of corrosion products of copper at one year of exposition: (a) Auger map and (b) AES in Tijuana (2010).

Figures 6 and 7 show the Auger spectra at 1, 3 and 6 months of metallic coupons, after the cleaning process with Ar^+ ions, at 5 minutes. SEM and AES analyses were carried out to determine the corrosion products formed in the copper surface. Figures 6 and 7show the scanning electron micrograph (SEM) images of areas selected for AES analysis covered by the principal corrosion products which are rich in the both air pollutants mentioned above in both cities as the test results indicate. The Auger map process was performed to analyze punctual zones, indicating the presence of Cl^- and $S^=$ as the main corrosive ions present in the copper corrosion products. The Auger spectra of Cu specimens was generated using a 5keV electron beam, which shows an analysis the chemical composition of the thin films formed in the Cu surface in Ensenada (Fig 2) and in Tijuana (Fig 3). The AES spectra of copper specimens installed in industrial plants in both cities show the surface analysis of three points evaluated in different zones of the metallic probes. The peaks of Cu appear between 905 and 915 eV, finding the chlorides and sulphides. In figure 6, the spectra reveals the presence of carbon and oxygen, chlorides and sulphides, with variable concentration in the chemical composition in the three regions analyzed, where the principal pollutant was Cl^- ion.

	Ensenada			TIJUANA		
Elements	Point 1	Point 2	Point 3	Point 1	Point 2	Point 3
C	36	29	29	24	26	23
Cl	12	32	31	X	X	X
Cu	13	11	12	19	21	26
O	8	12	13	26	23	19
S	31	16	15	31	30	32

Table 4. Atomic concentration (%) of metallic probes.

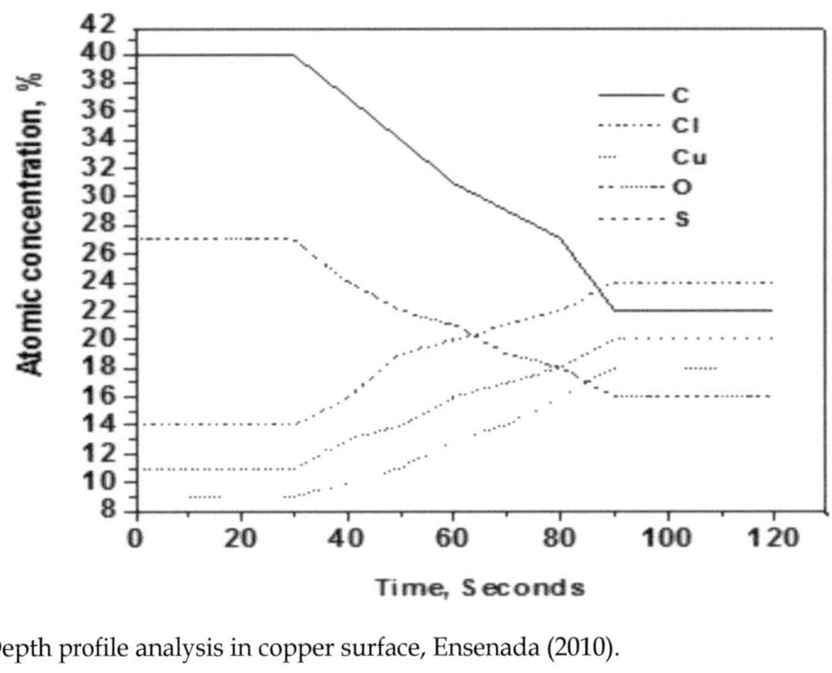

Fig. 6. Depth profile analysis in copper surface, Ensenada (2010).

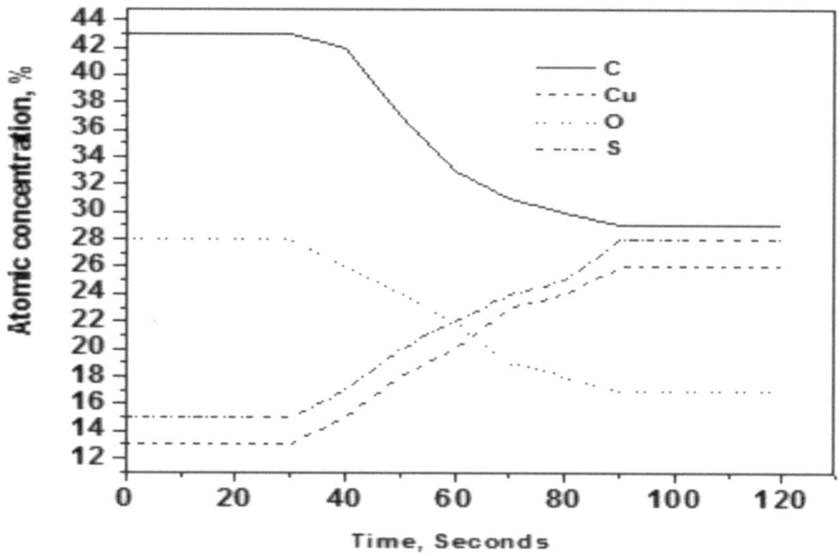

Fig. 7. Depth profile analysis in copper surface, Mexicali (2010).

The spectra reveal the importance of technical analysis with the Auger, which is evaluated with the formation of the films formed on copper surface thereby know the mechanism of corrosion in this metal. Auger depth profiles were collected on specimens of both cities, showed in figures 6 and 7. The depth profiling technique is defined by alternating cycles of Ar^{+}-ion sputtering to remove a thin layer (5 to 10 Å) of air pollutants that react with the copper surface and their characterization in some regions with the AES technique. In figures 6 and 7, chloride and sulfur located between carbide particles sputtered completely off during the first sputtering cycle (10 Å). A small chloride and sulphide persisted deeper into the carbide particles (point 2). In figure 6, the depth profile indicates a small presence of sulfur between the carbide particles.

4. Discussion

The air pollutants affect the deterioration of copper and its corrosion behavior and resistance. The principal anthropogenic and natural sources in indoor of industrial plants in both cities of corrosive pollutants are the gas emissions of vehicles, chloride particles from the marine environment and sulfides from the thermoelectrical power plants in Tijuana, Rosarito and Ensenada. Generation of corrosion in industrial plants has been an important factor in the last 30 years by the complexity of the electro-electronic devices and equipments that are qualified by the market demand, their operation and reliability (Lopez et al, 2007). The competition is governed by manufacture of electronic devices and equipments, increasing the necessity of develop various and great quantity of operations and decrease their size at a low cost (Valdez at al, 2006). This been has the principal effect to change the designs with smaller spaces between electronic devices and the use of new materials in electrical connectors and connections of electronic devices and equipments. Other factors are the uncontrolled climates of indoors, that promote the generation of corrosion (Lopez, 2008).

In most cases companies do not know the phenomenon of corrosion or not are considered as an important factor, until it causes a failure in some electronic devices and equipments, and stop the manufacture process.

5. Conclusion

Miniaturization and the requirement for high component density of small electronic devices need closer spacing and thinner metallic paths affected by corrosion and electrical failures. Copper exposed to air pollutants reveal that an increase on their concentrations at outdoor conditions has a critical impact on the indoor corrosion process in arid and marine environments. RH values higher than 75 % and concentration of air pollutants, promotes corrosion. The composition of the copper surface was obtained by the Auger spectra, showing localized corrosion from the first month to six months of exposure in both cities, and uniform corrosion. Particulate and gaseous pollutants, deposited on metal surfaces of micro-electronic components, are generated at residential and industrial zones with high motor traffic and operations in warehouses and offices, which promote corrosion. Electronic equipment installed in industrial plants are exposed to environmental factors in indoor and outdoor conditions. The corrosion of copper in indoor environments may be viewed as a variation of outdoor atmospheric corrosion. In contrast to outdoor exposure, in an indoor environment the wet film on the metal surface is thinner and it is often governed by relatively constant controlled humidity conditions. Sometimes the indoor environment temperature and RH are controlled and as a consequence, the amount of adsorbed water on surfaces is minimal and is constrained within reasonably tight limits. Currently measuring equipment such surface analysis techniques as AES was used in most of the industrial processes are very used to detect particles added to the metallic surfaces. In this study, Auger spectroscopy was made to detect the principal components added to surface of electrical connections and connectors. With this techniques, can obtain results of the chemical reaction the atmospheric agents that forms the thin films in metals of copper. Miniaturization and the requirement for high component density of small electronic devices need closer spacing and thinner metallic paths that originates the corrosion phenomena and electrical failures in the connections. Uniform and localized corrosion mechanisms are detected in electronic systems. Particulate and gaseous pollutants deposited on metal surfaces of micro-electronic components generated by traffic vehicles and operations of thermo electrical located at 50 kms of each city and provide electricity to this region in warehouses and offices, and promote corrosion. Electronic equipments installed in industrial plants are exposed to environmental factors, indoor and outdoor.

6. Acknowledgments

The authors express their gratitude for the financial support, of a Postdoctoral Scholarship to Gustavo Lopez by the Consejo Nacional de Ciencia y Tecnologia, trough Centro de Investigacion y de Educacion Superior de Ensenada and Universidad Nacional Autonoma de Mexico in Ensenada.

7. References

Annual Book of ASTM Standards, 2000, Wear and Erosion: Metal Corrosion, Vol. 03.02.

ASHRAE; Handbook; Heating, Ventilating and Ari-Conditioning; applications; *American Society of Heating*, Refrigerating and Air-Conditioning Engineers Inc.; 1999.

Asami K., Kikuchi M. and Hashimoto K.; An auger electron spectroscopic study of the corrosion behavior of an amorphous $Zr_{40}Cu_{60}$ alloy; *Corrosion Science;* Volume 39, Issue 1, January 1997, Pages 95-106; 1997.

Briggs D. and Seah M. P., Practical surface analysis, Second Edition, Volume 1 Auger and XPS, Photoelectron Spectroscopy,1990.

Clark A. E., Pantan C. G, Hench L. L; Auger Spectroscopic Analysis of Bioglass Corrosion Films; *Journal of the American Ceramic Society;* Volume 59 Issue 1-2, Pages 37–39; 2006.

Cole S. and Paterson D. A.; Relation of atmospheric pollution and the generation of corrosion in metals of copper, steel and nickel; Corrosion Engineering; 2004.

Consejo Nacional de Poblacion (CNP), Anuario Estadistico, Censo de Poblacion, Gobierno de Mexico, 2010.

Dillon P., MTI & DOE Launch Project Partnerships, Communications Materials Technology Institute of the Chemical Process Industries, Inc, 2000.

Duncan Balachandran, Walsh Harold; An Engineer's Guide to MATLAB, 2e: with Applications Electrical Systems, ; Prentice Hall, 2005. ISO 9223:1992, Corrosion of metals and alloys, Corrosivity of Atmospheres, Classification. ISO 11844-2:2005. Corrosion of metals and alloys - Classification of low corrosivity of indoor atmospheres - Determination and estimation attack in indoor atmospheres. ISO, Geneva, 2005. ISO 11844-1:2006. Corrosion of metals and alloys - Classification of low corrosivity of indoor atmospheres- Determination and estimation of indoor corrosivity. ISO, Geneva, 2006.

Lopez B.G.; Ph.D. Thesis; Caracterizaci6n de la corrosi6n en materiales metálicos de la industria electr6nica en Mexicali, B.C., 2008 (Spanish).

Lopez B.G., Valdez S.B., Zlatev K.R., Flores P.J., Carrillo B.M. and Schorr W. M.; Corrosion *of metals at* indoor conditions in the electronics manufacturing industry; *Anti-Corrosion Methods and Materials;* 2007.

Lopez B. G., Valdez S. B., Schorr W. M., Tiznado V. H., Soto H. G., Influence of climate factors on copper corrosion in electronic equipments and devices, *Anti-Corrosion Methods and Materials;* 2010.

Lopez G., Tiznado H., Soto G., De la Cruz W., Valdez B., Schorr M., Zlatev R.; "Corrosion de dispositivos electronicos por contaminacion atmosferica en intertiores de plantas de ambientes aridos y marinos; *Nova Scientia,* ISSN 2007-0705; 2010 (in Spanish).

Lopez B. Gustavo, Valdez S. Benjamin, Schorr W. Miguel, Zlatev R., Tiznado V. Hugo, Soto H. Gerardo, De la Cruz W.; AES in corrosion of electronic devices in arid in marine environments; *AntiCorrosion Methods and Materials;* (in press).

Moncmanova A. Ed. ; Environmental Deterioration of Materials, WITPress, 2007, pp 108-112.

Sankara Narayanan, Young Woo Park and Kang Yong Lee, Science direct, Elsevier B.V, "Fretting-corrosion mapping of tin-plated copper alloy contacts", Volume 262, Issues 1-2, 4 January 2007, Pag 228-233.

Swart H.C., Terblans J.J., Coetsee E., Kumar V., Ntwaeaborwa O.M.,Dhlamini M.S., Dolo J.J., Auger electron spectroscopy and X-ray photoelectron spectroscopy study of the electron-stimulated surface chemical reaction mechanism for phosphor degradation, Surface and Interface Analysis, Accesed: http://www3.interscience.wiley.com/journal/123317463/abstract, 2010.

Traviña A., Ortiz-Figueroa M., Cosio M.; Santa Ana winds and upwelling filaments off the Northern Baja California winds; Journal of Dynamic of Atmospheres and Oceans; 2002 .

Valdez B. and Schorr M.; El control de la corrosión en la industria electrónica; *Revista Ciencia*; 2006 (Spanish).

Veleva L., Valdez B., Lopez G., Vargas L. and Flores J.; Atmospheric corrosion of electro-electronics metals in urban desert simulated indoor environment; *Corrosion Engineering Science and Technology*; 2008.

Yves Van Ingelgem *, Isabelle Vandendael, Jean Vereecken, Annick Hubin, Study of copper corrosion products formed during localized corrosion using field emission Auger electron spectroscopy, *Surface and Interface Analysis*, Volume 40 Issue 3-4, Pages 273 -276, 2008.

Zlatev R., Valdez B., Stoycheva M., Vargas L., Lopez G., Schorr M.; Simpsoium 16: NACE "Corrosion and Metallurgy"; IMRC 2009, Cancun, Mexico.

Part 2

Indoor Air Quality

5

Correlation of Professional Performance to Acceptable IAQ in Critical Care Medical Facilities

H.W. Holder[1], K.V. Easterwood, Jr.[2], D.E. Johnson[3], J.W. Sealy[4], M.D.Larranaga[5] and D.C. Straus[6]

[1]SWK LLC
[2]Legacy Consulting, LLC
[3]SWK LLC
[4]Jim Sealy, Architect Consultant
[5]Oklahoma State University
[6]Texas Tech University Health Sciences Center
USA

1. Introduction

Fundamentally, buildings are simple things. Its basic purpose is to provide shelter. Initially, caves, cow hides for tents, hay for roofs, and mud for walls fulfilled this function. As life became more complex, beyond looking for the next meal, buildings followed suit. Historically, humanity has evolved from utilizing natural materials and living with the inherent limitations of these materials to integrating manufactured products limited only by the imagination of the designer. But if form truly "follows function"[1] there is possibly no more of a complex and critical function than that of our critical care medical facilities. These functions are so critical that, for the most part, a team of specialists is required to provide a fully operational facility. The realization of a building from a building program (written statement of need) into a three dimensional form is typically "chaired" by the architect who similarly employs the talents of various engineers and other specialists. All of this talent demands, and deserves, a fee structure higher than what is typically seen for office and other commercial buildings. Yet, even with the skill levels in place to bring such a complex environment together, flawed and defective buildings are designed and constructed. The ramifications of unsuccessful hospitals impact the very purpose of its mission and impose unnecessary burden on staff and management, but more significantly: **Defective Buildings shorten the life of buildings and defective critical care medical buildings present serious health and safety risks to patients and staff.**

---

[1] "It is the pervading law of all things organic, and inorganic, of all things physical and metaphysical, of all things human and all things super-human, of all true manifestations of the head, of the heart, of the soul, that the life is recognizable in its expression, that **form ever follows function**. This is the law. "American architect, Louis Sullivan, (1856-1924)

Various sub-systems that comprise the whole building (thus making the building a system) include, but are not limited to, the roof, walls and fenestration, heating, ventilation and air-conditioning system (HVAC), plumbing, electrical, finishes, furnishings, equipment and communication systems. At present, the quality of the indoor environment for the health and safety of occupants and protection of assets is becoming a prime concern with end users. Prior to the 1970's, buildings were built with natural and conventional materials. Designers and constructors were well experienced and trained in working with those materials and systems that were straight forward in concept. In the early 1970's, an oil shortage created the need to design and build more energy efficient buildings. As a result, ventilation requirements were substantially reduced in the interest of saving energy. That decision proved to be problematic.

During the 1980's building boom in the United States, the goal shifted to building cheaply and quickly, rather than building with the care seen in previous decades. Combined with the dictates of the 1970's dealing with energy adverse human response to the built environment was becoming evident and the term sick building syndrome (SBS) was coined by the World Health Organization in 1982. In the 1990's, the ventilation standard was revised to increase ventilation to address the SBS issue, as it was believed that dilution would be the solution (American Society of Heating Refrigeration and Air Conditioning Engineers [ASHRAE], ANSI/ASHRAE Standard 62-1989 Ventilation for Indoor Air Quality). Building codes and standards of care were originally developed for basically one reason, public health and safety. As the building industry evolved other codes such as the energy codes have joined the family of public safety codes. However, codes set minimum requirements and thus cannot guarantee quality, longevity or a *healthful* indoor environment. The combination of cheaper built buildings with increased ventilation introduced unplanned for moisture into the indoor environment, and later, the understanding of the role of microorganisms into the SBS vernacular, especially in warm and humid climates (Cooley et al., 1988). In the 2000's, buildings became more complicated due to the need for specific functional use, rapidly changing technology, and the creative application of both conventional and newly developed composite and synthetic materials. Furthermore, the building contractor became more of a broker than a builder due to the economics of tight schedules and budget driven contracts. Today's hospital environment requires a healthcare facility's HVAC systems to provide excellent ventilation effectiveness in order to maintain appropriate indoor air quality, prevent the spread of infection, preserve a sterile and healing environment for patients and staff and to maintain space and comfort conditions. These demands require a healthcare facility's HVAC systems to provide significant quantities of total ventilation and outdoor air. They also require significant treatment of this ventilation air, including cooling, dehumidifying, reheating, humidifying, and filtration of the air to achieve these effective ventilation goals. Trends indicate that even more treatment of the air will be required to respond to infection control and bioterrorism issues in the future.

Given the evolution of the design and construction industry and the diminished quality of construction due to a steeply declining skill set in the building trades, the useful life expectancy of a building, other than strictly controlled construction for institutional buildings.[2] may no longer be 50-60 years, but significantly less.[3]

---

[2] The case studies presented here are small, regional facilities that lacked the staff and oversight typically seen in large university hospitals or similar facilities.

2. Contributing causes to a defective building

The following factors have played a part in our findings in addition to inadequate budgets and overly optimistic building programs.

2.1 The economy v. mentoring system

Most design professions require an advanced or professional degree at a recognized institution of higher learning. Internship requirements follow a professional education (three years for aspiring architects and engineers). Finally, a professional exam issued by a state is required for licensure. In the past, aspiring design professionals had the opportunity to sit by the side of the "master" and learn the craft of designing a building and its systems. Inflation and the demands of a super-heated economy have changed this scenario. Today, when everything is available instantly, and time is money, the opportunity to engage and learn from the "master" is long gone and this void is worsened by;

2.2 Computers v. knowledge

Although the use of Computer Aided Design systems (CAD) is now universal and certainly can be a great asset to the design team, detail "libraries" can subvert the process of thinking through how things fit together and perform within an overall system. The authors have reviewed construction documents that contain pages and pages of details with the notation that they "may not be a part of this project".[4] This language is completely unacceptable, but does support the notion that some architects either do not understand the importance of a clear set of construction documents, or are yielding to time constraints.

2.3 Specialization itself

The myriad complexities of the typical critical care medical facility require a significant amount of time in coordination with the end user and various disciplines required to ensure the efficient functioning of this facility. For example, Building codes, accessibility standards, and energy requirements must be addressed, as well as state health codes and the Center for Disease Control ("CDC") infectious disease control guidelines (CDC 2003). The authors' review determined 1) the architectural design team was so focused on the functional complexities that the fundamental task of ensuring that the building envelop would keep out the elements was overlooked; 2) the construction documents and or industry standards were not followed resulting in system failures; 3) the fundamentals of mechanical design for the climatic location of the building and or operating standards were not appropriately applied, and 4) the environmental systems were poorly installed, were not sufficiently commissioned, and did not perform to minimum standards.

3. Case studies

The following case studies involved architects that were ostensibly seasoned and who had specialized in the design of critical care medical facilities.

---

[3] Currently unknown as to unintended consequences "Green" or "Sustainable Buildings" associated with the U.S. Green Building Council's Leed Certifications have recently been developed to address energy issues and diminishing resources.

[4] Found on the construction document for a multi-million dollar renovation of a high school in Texas, circa 2005.

3.1 Case study 1("CS 1")

This single story surgery center was constructed in 2004. It was 19,000 square feet and constructed under a design/build delivery system.[5] Its exterior was composed of applied stone veneer and an Exterior Finish Insulation System (EFIS). The hospital staff noticed mold growth, wild swings in temperature, and wet indoor environment almost from the day they occupied the building. An investigation ensued and determined that defects discovered associated with the shell of the building, or envelope, was a major contributor, along with the HVAC, to the difficulties they were encountering. Site drainage violated code and held water against the building. It is fundamental to the proper placement of a building to locate it on the site both vertically and horizontally. The horizontal aspect of the placement is guided by local zoning codes. Vertical placement is guided by common sense, water should drain away from a building and thus, downhill. Building codes have language to ensure that the common sense approach is met. For example the building code reads: "The ground immediately adjacent to the foundation shall be sloped away from the building at a slope of not less than one unit vertical in 20 units horizontal (5-percent slope) for a minimum distance of 10 feet (3048 mm) measured perpendicular to the face of the wall or an approved alternate method of diverting water away from the foundation shall be used." (International Building Code [IBC] Chapter 18: Soils and Foundation, 1803.3 Site Grading, 2000).

Compounding this hospital's flawed start were flashings that were either missing or misapplied. It follows that if it is important to drain water away from a building, it is just as important to drain it out of the exterior wall systems. To accomplish this, drainage planes, flashings and weeps should be well known by architects, engineers and constructors and they are well documented in industry and professional publications. Codes, again, have language to ensure these fundamentals are met. For example, the 2000 Edition of the International Building Code reads: "Exterior walls shall provide the building with a weather-resistant exterior wall envelope. The exterior wall envelope shall include flashing, as described in Section 1405.3. The exterior wall envelope shall be designed and constructed in such a manner as to prevent the accumulation of water within the wall assembly by providing a water-resistive barrier behind the exterior veneer, as described in Section 1402.2 and a means for draining water that enters the assembly to the exterior of the veneer, unless it is determined that penetration of water behind the veneer shall not be detrimental to the building performance. Protection against condensation in the exterior wall assemble shall be provide in accordance with the International Energy Conservation Code" (IBC, Exterior Walls, Section 1403.2, Weather Protection, 2000).

When combined with the fact that water stands against the building, the base flashing not only did not allow water to drain it instead reversed the flow of water such that outside water entered the building. The end result was mold growth along the entire perimeter of the building.

Doors, windows, and louvers all penetrate the walls, thus flashings are also used to divert water to the exterior of a building. Codes (IBC Section 1405.3; Flashing., 2003) and manufacturer's details leave no doubt that the flashing of openings is a requirement, and that information as to materials and methods are readily available. The consequence of ignorance or neglect is water pouring into the building.

---

[5] Design/Build is where the builder contracts with the owner to design and construct a project as opposed to the traditional method where the owner hires an architect to design and a builder to construct.

Fig. 1. Site drainage violated code and, held water against the building.

Fig. 2. The copper-coated base flashing at the applied masonry veneer wall terminated below grade and therefore did not allow for the water to drain.

Although flashings are addressed in building codes (IBC Section 1405.3; Flashing , 2003), it is more perplexing as there were a myriad of opportunities for the CS 1 architect and contractor to determine the appropriate method to terminate the exterior wall. Industry organizations, product manufacturers and associations all have literature readily available on the internet.[6] This information is not only available but details and specifications can be downloaded for use by the architect.[7] Furthermore the General Conditions of the Contract for Construction Standard Form of Agreement A201 [American Institute of Architects (AIA), 1997] delineate processes to resolve conflicts. Above the veneered surface of the CS 1 exterior wall, and continuing to the roof was EFIS. EFIS has become widely popular as a light weight, inexpensive system that has the appearance of stucco. Early on, problems arose regarding this system's propensity to hold water between the insulation board and the exterior sheathing which led to mold growth and eventually, numerous law suits.[8] The industry responded by improving the product and producing numerous details and installation procedures, all of which can be accessed via the internet.[9]

Fig. 3. Mold caused by moisture in the wall cavity.

---

[6] For example, The Rocky Mountain Masonry Institute, Denver, CO; The Brick Industry Association, Reston VA; The Portland Cement Association, Skokie, IL.

[7] For example details and specifications regarding the detailing of masonry can be found at The Brick Institute of America's website: www.gobrick.com/TechinicalNotes/tabid/7658/Default.aspx (last visited on February 13, 2011)

[8] The most notable being the Marin County Courthouse, CA.

[9]http://www.dryvit.com/specs.asp?country_id=1. (last visited on February 13, 2011)

Fig. 4. Water entering the building because of inadequate flashing.

Negative building pressure and improper pressure relationships were the major contributor to the poor physical environment. Pressure relationships between critical care areas inside the building with respect to other spaces were not in accordance with [Title 25, Texas Administrative Code, Chapter 135, Ambulatory Surgical Centers Licensing Rules, (ASCLR), 2009]. Both design and installation defects contributed to these conditions. The design of the HVAC system was inadequate for humidity control. A test, adjust, balance, and commissioning process was obviously never completely performed for the performance aspects of the project other than the air volume measurements. The HVAC control system was deficient in that it was never commissioned to control to the various pieces of equipment as an integrated system.

In a response to uncontrolled high indoor humidity, areas were over-cooled including operating rooms (OR), which resulted in frequent condensation on the interior surfaces, medical equipment, and supplies to the extent surgeries had to be rescheduled or cancelled. Water stained building materials, indoor mold growth, temperature fluctuations, high humidity, and condensation all contributed to an unsatisfactory environment for patients and staff and *violated state licensing rules.* Roof top air handling unit panels for access to the internal equipment and penetrations for conduit, refrigerant lines, etc. leaked water, air, and dust into the interior of the units. Water intrusion stains, mold growth, and debris were found in the air handling units and on ceilings below the air handling units. This project was a classic example of poor coordination. As examples: 1) it is common knowledge that vinyl wall coverings in a hot and humid climate are doomed to failure, yet vinyl wall coverings were specified and installed; 2) engineers know that buildings located in hot and humid climates need HVAC systems designed to work in that climate, yet only the dry-bulb/coincident wet-bulb design condition was used for coil selections resulting in high humidity indoors; and 3) the contractor and subcontractors not only performed substandard and incorrect work, they ignored the consultants' notice of such. Both the architect and contractor failed to provide the quality control and oversight required to deliver an acceptable project. The architect 1) failed to confirm that the site design conformed to code and common sense; 2) failed to detail the flashings to prevent water intrusion, and 3) failed to observe that the weather barrier behind the EFIS was missing. The contractor failed to notify the architect of this omission. The information regarding all of these discovered defects was readily available. The defects caused the building materials to be wetted for prolonged periods of time resulting in mold growth in walls, behind millwork and on ceilings. In short, this project failed "Design and Construction 101" and corrections that should have been made in the field are now to be made in the courtroom.

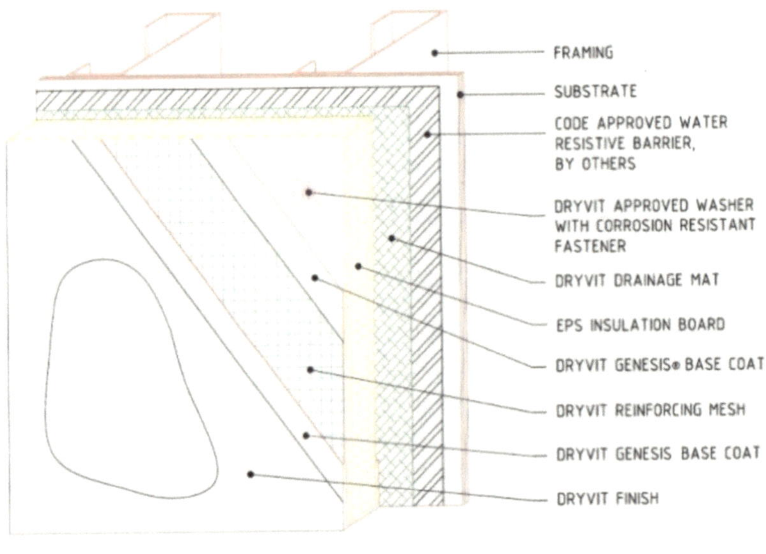

Fig. 5. Typical detail.

Rule #	Subject	Violation (rule language)
§135.52.(h)(5)	HVAC	All rooms and areas in the ASC shall have provision for positive ventilation. Fans serving exhaust systems shall be located at the discharge end and shall be conveniently accessible for service. Exhaust systems may be combined, unless otherwise noted, for efficient use of recovery devices required for energy conservation. The ventilation rates shown in Table 1 of §135.56(a) of this title shall be used only as minimum requirements, since they do not preclude the use of higher rates that may be appropriate.

Rule #	Subject	Violation (rule language)
(C)	General ventilation requirements	All rooms and areas in the ASC shall have provision for positive ventilation. The ventilation rates shown in Table 1 of §135.56(a) of this title shall be used only as minimum requirements, since they do not preclude the use of higher rates that may be appropriate.
(C)(iv)	Temperatures and humidities	The designed capacity of the systems shall be capable of providing the ranges of temperatures and humidities as shown in Table 1 of §135.56(a) of this title.
(C)(viii)	Directional air flow	Ventilation systems shall be designed and balanced to provide pressure relationships contained in Table 1 of §135.56(a) of this title.
(C)(x)	Ventilation start-up requirements	Air handling systems shall not be started or operated without the filters installed in place. This includes the 90% and 99.97% efficiency filters where required. This includes during construction operations. Ducts shall be cleaned thoroughly and throughout by a National Air Duct Cleaners Association (NADCA) certified air duct cleaning contractor when the air handling systems have been operating without the required filters in place. When ducts are determined to be dirty or dusty, the department shall require a written report assuring cleanliness of duct and clean air quality.
(C)(xi)	Humidifier location	When duct humidifiers are located upstream of the final filters, they shall be located at least 15 feet from the filters. Duct work with duct-mounted humidifiers shall be provided with a means of removing water accumulation. An adjustable high-limit humidistat shall be located downstream of the humidifier to reduce the potential of condensation inside the duct.
(C)(xii)(V)	Pressure monitoring devices	A manometer or draft gauge shall be installed across each filter bed having a required efficiency of 75% or more, including laboratory hoods requiring high efficiency particulate air (HEPA) filters. The pressure monitoring device shall be mounted below the ceiling line within the ASC such that it can be observed by staff.
(H)	Fire damper requirements	Fire dampers shall be located and installed in all ducts at the point of penetration of a required two-hour or higher fire-rated wall or floor in accordance with the requirements of NFPA 101, §18.5.2.
(I)	Smoke damper requirements	Smoke dampers shall be located and installed in accordance with the requirements of NFPA 101, §20 .3.7.3, and NFPA 90A, Chapter 5.
(M)	Make-up air	If air supply requirements in Table 2 of §135.56(b) of this title do not provide sufficient air for use by exhaust hoods and safety cabinets, filtered make-up air shall be ducted to maintain the required air flow direction in that room.

Table 1. Nine of twenty-four rules in ASCLR for HVAC systems were noncompliant.

3.2 Case study 2 ("CS 2")

This single story hospital was constructed in 2006. It was 33,000 square feet and constructed under a design/build delivery system. Its exterior was composed of brick veneer with a modified bitumen roof over a steel roof structure. The hospital staff noticed mold growth, wild swings in temperature, and wet indoor environment almost from the day they occupied the building. An investigation ensued and determined that defects discovered associated with the shell of the building, or envelope, was a major contributor, along with the HVAC to the difficulties they were encountering.

Fig. 6. Site drainage violated code and held water against the building.

As in CS 1, common sense and conformance to building codes were not observed at this project. For example, fig. 7 showed water standing at a door that subsequently entered the building.

Fig. 7. A flawed detail at the base of the building compounded the threat to the building posed by standing water.

A standard detail that would allow a veneer to terminate below grade calls for a flashing to be installed at a brick or masonry course above grade. The void below the flashing is then filled with grout that, essentially, makes the veneer below grade a part of the foundation.

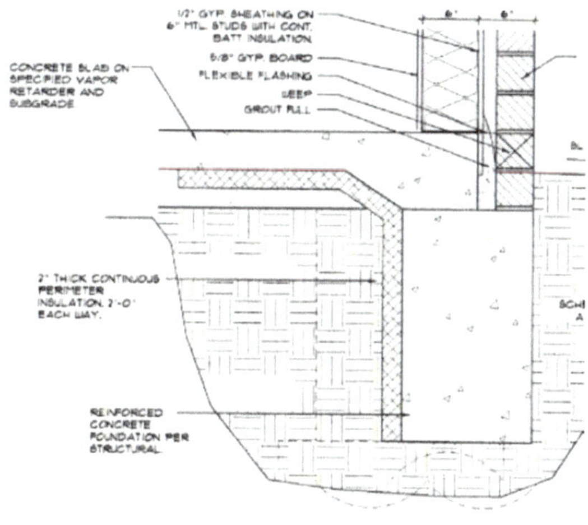

Fig. 8. The architect's detail correctly shows the acceptable method of installing base flashing when the masonry terminates into grade.

However, he failed to follow manufacturer's recommendations and buried the exterior sheathing into the grout fill. Inattention to detail by both architect, who failed to refer to the manufacturer's guidelines, and contractor, who failed to notice the defect, led to the unacceptable condition. As a result of these errors water stands within the wall cavity posing a continuing threat to building materials.

Fig. 9. Water standing inside the wall cavity.

Roof coverings have evolved over time allowing several systems from which to choose. All of these systems have advantages and disadvantages. Depending on the selection, warranties are available for up to twenty-five years.[10] Manufacturer's standard details are typically incorporated into the construction set of drawings. Additionally, there is typically a requirement within the construction documents (drawings and specifications) for the roof manufacturer to be present during certain phases of the work, including a pre-construction meeting. Even so, roofs fail as do roofing contractors. So it becomes essential that the details are followed by the roofer and the installation is closely scrutinized by third-parties. Perhaps the biggest challenge in accomplishing a sound roof rests in the details that deal with roof penetrations by pipes, vents and ducts. In this hospital, the architect and mechanical engineer provided details to ensure the roof would be sound. Roof details were included in the construction documents by the architect and the mechanical engineer to accommodate thermal movement that can result in tearing of the roof membrane, and to prevent water from entering the building. These details were ignored, resulting in water damage to materials and, of critical importance, added moisture into the building that exacerbated the already poor environmental condition within the hospital.

The CS 2 facility is under assault from above and below. Water entered the building from the site that was poorly graded and was retained in the wall system by a detail that was poorly executed. As in CS 1, adherence to common sense should have dictated a positive flow of water away from the structure. The architect was not retained to perform Construction Administration Services ("CA"), thus severing a key element of quality control during the construction process. "Full service" would have called for checks and balances as dictated by the American Institute of Architects Contract Documents A201, General Conditions and supported by AIA Documents G711 (AIA, 1972), 712 (AIA, 1972), 714 (AIA, 2007) and 716 (AIA, 2004) for use by both the contractor and architect. By not having CA in place, the contractor implies that all elements of the project conform to the Contract Documents (plans and specifications), as well as building codes, which is not the case here.

Fig. 10. Water running out of insulation.

---

10 http://www.specjm.com/commercial/roofing/peakadvantageguarantees.asp (last visited on February 13, 2011)

Fig. 11. Improper site grading.

A value engineering (VE) exercise resulted in the use of direct expansion (DX) roof top units with coils that had a sensible heat ratio too high to effectively dehumidify. Exhaust fans were VE and as a result the fans were located below the roof which was a violation of [Title 25, Texas Administrative Code, Chapter 133, Hospital Licensing State Regulations (HLSR) 2007]. It was obvious that the VE items selected were money driven and health and safety concerns and/or the HLSR rules were ignored. Without an architects CA services in place to mind the store and with an unsophisticated owner, the contractor and his subcontractors did whatever they wanted to do. Negative building pressure and improper pressure relationships were the major contributor to the poor physical environment. Pressure relationships between critical care areas inside the building with respect to other spaces were not in accordance with HLSR. Both design and installation defects contributed to these conditions. The design of the HVAC system was inadequate for humidity control. Engineers know that buildings located in hot and humid climates need HVAC systems designed to work in that climate yet only the dry-bulb/coincident wet-bulb design condition was considered for coil selections. A test, adjust, balance, and commissioning process was obviously never completely performed for the performance aspects of the project other than the air volume measurements. The HVAC control system was deficient in that it was never commissioned to control to the various pieces of equipment as an integrated system. In fact the maintenance staff at the facility was never trained to operate the system and they had to resort to an on-line third party provider for day-to-day control.

In a response to uncontrolled high indoor humidity, areas were over-cooled including OR, which resulted in frequent condensation on the interior surfaces, medical equipment, and supplies to the extent surgeries had to be rescheduled or cancelled. Water stained building materials, indoor mold growth, temperature fluctuations, high humidity, and condensation all contributed to an unsatisfactory environment for patients and staff and *violated state licensing rules.*

Rule #	Subject HVAC	Violation (rule language)
§133.162.(d)(3)	General Mechanical Requirements	All rooms and areas in the ASC shall have provision for positive ventilation. The ventilation rates shown in Table 1 of §135.56(a) of this title shall be used only as minimum requirements, since they do not preclude the use of higher rates that may be appropriate.
(C)	Performance and acceptance	Prior to completion and acceptance of the facility, all mechanical systems shall be tested, balanced, and operated to demonstrate to the design engineer or his representative that the installation and performance of these systems conform to the requirements of the plans and specifications.
(C)(i)	Material lists	Upon completion of the contract, the owner shall be provided with parts lists and procurement information with numbers and description for each piece of equipment.
(D)(i)(III)	Outside air location	Outside air intakes shall be located at least 25 feet from exhaust outlets of ventilating systems, combustion equipment stacks, medical- surgical vacuum systems, plumbing vents, or areas which may collect vehicular exhaust or other noxious fumes. (Prevailing winds and proximity to other structures may require more stringent requirements). Plumbing and vacuum vents that terminate five feet above the level of the top of the air intake may be located as close as 10 feet.
(D)(i)(VI)	Directional air flow	Ventilation systems shall be designed and balanced to provide directional flow as shown in Table 3 of §133.169(c) of this title. For reductions and shutdown of ventilation systems when a room is unoccupied, the provisions in Note 4 of Table 3 of §133.169(c) of this title shall be followed.
(D)(i)(IX)	Humidifier location	When duct humidifiers are located upstream of the final filters, they shall be located at least 15 feet from the filters. Ductwork with duct- mounted humidifiers shall be provided with a means of removing water accumulation. An adjustable high-limit humidistat shall be located downstream of the humidifier to reduce the potential of condensation inside the duct.
(D)(iv)(II)	Smoke removal systems	Smoke removal systems for surgical suites. Smoke removal systems shall be provided in all surgical suites in accordance with NFPA 99, §6.4.1.3.
(D)(vii)	Fire damper requirements	Fire dampers shall be located and installed in all ducts at the point of penetration of a required two-hour or higher fire rated wall or floor in accordance with the requirements of NFPA 101, §18.5.2.

Rule #	Subject HVAC	Violation (rule language)
(D)(viii)	Smoke damper requirements	Smoke dampers shall be located and installed in accordance with the requirements of NFPA 101, §18.3.7.3, and NFPA 90A, Chapter 5.
(D)(xii)	Make-up air	If air supply requirements in Table 3 of §133.169(c) of this title do not provide sufficient air for use by exhaust hoods and safety cabinets, filtered make-up air shall be ducted to maintain the required air flow direction in that room.
§133.162.(d)(4)	General Piping Systems and Plumbing Fixture Requirements	All piping systems and plumbing fixtures shall be designed and installed in accordance with the requirements of the National Standard Plumbing Code Illustrated published by the National Association of Plumbing-Heating-Cooling Contractors (PHCC), 2003 edition, and this paragraph.

Table 2. Ten of forty-four rules in HLSR for HVAC systems were noncompliant.

3.3 Case study 3 ("CS 3")

This single story hospital was constructed in 2009. It was 70,000 square feet and constructed under the more traditional design/bid/build delivery system. Its exterior was composed of concrete masonry units with accent bands of cast stone. In the previous two studies the issue of prime importance was water intrusion caused by site and envelope issues and the inability of the HVAC systems to control the environment. Water intrusion was an issue in this case, however, the prime concern was an unsecured envelope that allowed unacceptable levels of particulate (dust) into the hospital. Early on the hospital was notified by the State Licensing Agency that they must gain control of the dust.

As in the previous two case studies, the "broken record" continues as the CS 3 site does not sufficiently drain water away from the building. In CS 3, the civil engineer apparently was not aware of the code requirement that called for a 5% slope away from the building for a distance of ten feet. Additionally, the architect, who claimed to not have the responsibility for the Civil Engineer's work set the finish floor height. In a memo, the architect chided the engineer for delays and stated "…we have done most of the work anyway".[11]

Here, there are two professionals that do not have a working knowledge of the building code. Unit pricing for earthwork was relatively low compared to the overall budget for this building, and had the error been corrected at the outset, the cost to raise the building would be negligible to the overall budget. With a completed building that includes parking on four sides, sidewalks, and the complication of a very flat terrain, what should have cost a few thousand dollars to add fill under the slab was budgeted at close to a million dollars to rectify.

In spite of the lapse of coordination regarding the site, the architect provided numerous and correct details regarding the exterior of the building and its fenestration. The sections and details were supported by 1,162 pages of specifications. The architect had made 151 key references on the drawings alone that dealt with flashings and the job of directing water away from the building. These details were all ignored. As in CS 2 the flawed site and defective base condition caused water to stand within the exterior wall cavity.

[11] Memorandum from the Architect to the Civil Engineer obtained in pre-trial discovery.

The Architect's specifications and drawings were consistent and complementary. The contract with the general contractor (constructor) states that if the plans and specifications are in conflict, the most stringent requirement governs. In this case, the contractor saw fit to ignore both plans and specifications.

The CS 3 roof was also problematic and let particulate matter into the exterior walls and spaces above the ceilings. This hospital is located in an agricultural area, and dust is a continuing issue.

Fig. 12. Water standing inside wall cavity.

Figures 13, 14 and 15 show the progression of dust (particulate) from the parapet, into the exterior wall via mislapped flashing and finally into the ceiling space of the hospital. In agricultural settings dust is especially problematic in a critical care hospital. In a 2009 published study dust was found to contain endotoxins, allergens, microbial pathogens, bacterial toxins, fungal spores, and mycotoxins that can cause adverse human reactions. (Purdy & Straus 2009)

Fig. 13. Infiltration of dirt and moisture under parapet cap.

Fig. 14. Infiltration of dirt and moisture under parapet wall.

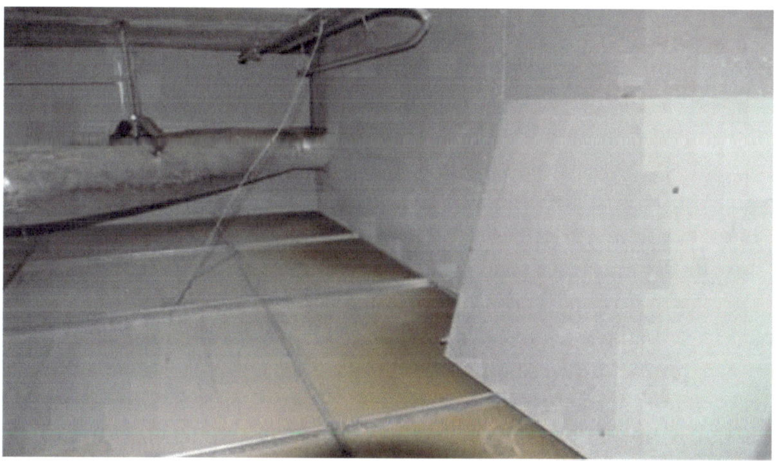

Fig. 15. Infiltration of dirt into the attic space above lay in ceiling.

With the exception of civil engineering, the quantity and quality of the architect's and structural engineer's documents (plans and specifications) were exceptionally good. Unlike CS 2, the architect here had a full service contract and attended regularly scheduled job site meetings. In spite of his presence at the job site, virtually all of his details regarding openings (windows and doors) were ignored by the contractor. The architect, who performed well on 80% of his contract with the owner will have to answer for the representative 20% of failed CA when he did not take the contractor to task for violating the contract for construction. This project was located in the southern high plains and one would think that humidity control would not be much of a concern. A chilled water HVAC system design was replaced in a VE exercise to direct expansion (DX) roof top units with coils that had a sensible heat ratio too high to effectively dehumidify during the relatively few hours of the year when dehumidification was required. This resulted in excursions of high indoor humidity where control of the indoor environment could not meet the HLSR requirements. Negative building pressure and improper pressure relationships were the major contributor to the poor physical environment. Pressure relationships between critical care areas inside the building with respect to other spaces were not in accordance with the HLSR. Both design and installation defects contributed to these conditions. The primary design defect of the HVAC system was the variable air volume (VAV) application with the VE DX cooling roof top units that need constant volume air flow. Frequent cycling of compressors resulted in temperature swings in the occupied space. The primary HVAC installation defect was excessive duct leakage which contributed to negative building pressure. However, there was a heating problem that no one seemed to be able to solve other than to reduce the amount of outside air to increase the discharge temperature, which increased negative building pressure. The mystery turned out to be a plumbing problem. The plumber located the natural gas pressure regulators in the medium pressure gas system too far from the unit and the pressure drop in the gas line starved the units for gas which reduced the capacity and did not heat the air effectively.

Excessive dust in the air is a common problem in the location of CS 3. Normal operation of the HVAC system filtered the outside air being introduced through the roof top units. During the frequent dust storms common to the area, the HVAC filters could be used up in one day. In an effort to deal with these frequent dust storms a "dust mode" control sequence was designed and installed into the HVAC control system. What seemed to be a good idea to someone turned out to be a disaster. Exhaust fans were not turned off in the dust mode because exhaust must be maintained in many areas (isolation rooms, certain critical care areas, etc.) of a hospital. Therefore, during dust mode the roof top units were not introducing outside air and the exhaust fans were sucking dust in from where ever it could find its way into the building. Even during ordinary operation (not dust mode) the building was operating in a negative pressure and sucking dust into the building creating an unsatisfactory environment for patients and staff.

A test, adjust, balance, and commissioning process was obviously never completely performed for the performance aspects of the project. Deficiencies were noted and sent to the engineer and contractor but were never followed-up. The HVAC control system was deficient in that it was never commissioned to control to the various pieces of equipment as an integrated system. In fact the maintenance staff at the facility was never trained to operate the system and they had to resort to an on-line third party provider for day-to-day control.

The building was under construction for a long time and the duct work was never protected during construction. The duct work was not cleaned before operating the equipment. Dust in the ductwork was deep enough to leave ruts when a robot was used to photograph the ductwork to an operating room.

Fig. 16. Tracks of robotic camera in the dust of the HVAC ductwork.

Rule #	Subject HVAC	Violation (rule language)
§133.162.(d)(3)	General Mechanical Requirements	All rooms and areas in the ASC shall have provision for positive ventilation. The ventilation rates shown in Table 1 of §135.56(a) of this title shall be used only as minimum requirements, since they do not preclude the use of higher rates that may be appropriate.
(C)	Performance and acceptance	Prior to completion and acceptance of the facility, all mechanical systems shall be tested, balanced, and operated to demonstrate to the design engineer or his representative that the installation and performance of these systems conform to the requirements of the plans and specifications.
(C)(i)	Material lists	Upon completion of the contract, the owner shall be provided with parts lists and procurement information with numbers and description for each piece of equipment.
(D)(i)(VI)	Directional air flow	Ventilation systems shall be designed and balanced to provide directional flow as shown in Table 3 of §133.169(c) of this title. For reductions and shutdown of ventilation systems when a room is unoccupied, the provisions in Note 4 of Table 3 of §133.169(c) of this title shall be followed.

(D)(i)(VIII)	Ventilation start-up requirements	Air handling systems shall not be started or operated without the filters installed in place. This includes the 90% and 99.97% efficiency filters where required. Ducts shall be cleaned thoroughly and throughout by a certified air duct cleaning contractor when the air handling systems have been operating without the required filters in place.
(D)(iv)(II)	Smoke removal systems	Smoke removal systems for surgical suites. Smoke removal systems shall be provided in all surgical suites in accordance with NFPA 99, §6.4.1.3.
(D)(ix)	Acceptable damper assemblies	Only fire damper and smoke damper assemblies integral with sleeves and listed for the intended purpose shall be acceptable.
(D)(xii)	Make-up air	If air supply requirements in Table 3 of §133.169(c) of this title do not provide sufficient air for use by exhaust hoods and safety cabinets, filtered make-up air shall be ducted to maintain the required air flow direction in that room.
§133.162.(d)(4)	General piping and plumbing	All piping systems and plumbing fixtures shall be designed and installed in accordance with the requirements of the National Standard Plumbing Code

Table 3. Nine of forty-four rules in HLSR for HVAC systems were noncompliant.

4. Common elements

4.1 Budget and cost savings

Discovered defects were not a result of cost savings or inadequate budgets. In the instances where the termination of the exterior wall was the issue (found in all three of these studies) the specified materials were in place although out of order, therefore the labor and material costs would be the same whether done correctly or incorrectly. The missing weather-resistive barrier of CS 1 does not represent more expense to the owner because no credit was given. In other words the contractor received money for materials and labor he did not provide. In every case the site grading and vertical placement of the building was problematic. Solutions and costs to bring the site into compliance and to remove the threat to the building were both complex and costly. A sixty thousand dollar increase in the original budget for CS 3 would have eliminated a nearly one million dollar repair.

4.2 Noncompliance with regulatory authority

Codes are minimum and not something to attain. Life safety is the basis for professional licensure. Numerous codes including the International Building Code, Plumbing, Electrical and Energy Codes are in place to protect the public. Non-compliance with code is inexcusable and can threaten life safety. Each design professional, contractors or sub-contractors is required to know the laws relating to their profession or avocation. Licenses are granted by the states and continuing education is mandatory. The excuse by a contractor or sub-contractor that they built what was drawn does not hold water.

All three case studies show that negative building pressure and improper pressure relationships between critical care areas with respect to other spaces was a, if not the, major contributor to a poor physical environment. A common design defect seen in most designs for medical facilities was an insufficient initial evaluation of fundamental design criteria for the HVAC systems cooling coils. In the above case studies it was obvious that the HVAC units were designed using dry bulb/coincident wet-bulb conditions and the wet bulb/coincident dry-bulb and the dew point/coincident dry-bulb conditions were not considered. In general, it appears to have been a business as usual approach, "what was done on the last job", when it came to criteria development, detail of design, equipment selection, and application decisions. The bottom line results were a flawed design that did not meet the letter or intent of minimum code and/or licensing rules.

Not evaluating all three design conditions for the location of the project can result in an uncontrolled humid indoor condition and the results can be ugly.

When DX HVAC systems end up on a hospital project either by design or VE, humidity control is severely compromised if not eliminated unless the latent cooling (dehumidification) is decoupled from the sensible cooling with dedicated outdoor air systems. DX HVAC systems that introduce outside air for ventilation and building pressurization directly into the air handler unit to mix with return air before entering the cooling coil are at a huge disadvantage when it comes to humidity control. The direct expansion refrigeration equipment used a hot gas by-pass and an "on/off" operation of compressors to control capacity. This resulted in broad temperature and humidity excursions in the conditioned space, condensation, and mold growth on surfaces. Designing successful hospital HVAC projects of any size is not rocket science. It is almost connect the dot technology for those who are willing to break away from the "way I did last time" modus operandi. ASHRAE has produced many guidelines (Harriman et al., 2001), standards of care (ASHRAE 2007), and training materials (ASHRAE 2003) for properly designing medical facilities. Decoupling the latent load from the sensible load with dedicated outside air systems (DOAS) is the most efficient and effective way to maintain an acceptable indoor environment, especially in a humid or moderate climate. Although it can be done other ways, those methods seem to get complicated for the normal staff of a small to regional sized facility.

Fig. 17. Mold under a nurses desk.

4.3 Poor contractor quality control

In the past building contractors "self performed" some of the work and sub-contracted out specialized trades. Increasingly contractors now take on more of the role of a "broker" and have less of an interest and relationship with the day to day operations of the construction site. Virtually all trades are now sub-contracted. The "broker" role tends to lessen the traditional responsibility of the contractor who, by contract, is to schedule, manage and coordinate the quality of the work and certify its conformance to the construction documents. Too often many projects are "run by the subs" meaning that the contractor is aloof to the daily operation of the job.

4.4 Inadequate or poor site observations by design professionals

In every case study the architect failed to observe major discrepancies with his construction documents. The contractor, as well, failed to notify the architect of possible discrepancies to be resolved. CS 3 was especially insightful and led to the conclusion that the complexities of a hospital's internal systems caused the architect to either delegate oversight of the building proper to an intern, assume that his details were followed without checking, or he forgot about it. In CS 3 the construction documents were exceptional yet virtually the entire exterior was flawed to the extent that it had to be replaced.

4.5 All critical care facilities receive co and license but does not mean it is OK

Because a building has periodic inspections by the Building Official after the initial review and issuance of a Building Permit and, later, a Certificate of Occupancy does not mean that defects are acceptable. Building officials, and other regulatory agencies, deal with entire communities and therefore must have some reliance on the professionals and their processes to provide a compliant building. Architects make periodic visits to a building site to view progress and to certify monthly that work is on schedule and the amount requested by the contractor is justified. The architect's periodic visits are not sufficient, and are not intended to be so thorough that all defects, misapplications or misinterpretations of his documents are observed. The building officials make far fewer site visits and the argument that they must have seen and approved of every element is hard to make with a straight face.

4.6 Value engineering

This is becoming industry short hand for "cheap". VE typically comes into play when cost overruns are encountered prior to the start of the construction itself. All parties participate-contractor, architect, major sub-contractors and the owner. Too often the major cost reduction will be in the HVAC system and inadequate systems are often offered as "savings". Furthermore, most VE offerings are several single systems, one being the HVAC, and little thought or engineering has gone into how VE changes effect the other systems and the building system as a whole. Although the HVAC system is a substantial part of the overall construction cost it must be the considered too vital to compromise with a VE exercise as it is obvious that even more treatment of ventilation air will be required to manage infection control and bioterrorism issues in the future. With a hospital, where infection control depends on ventilation effectiveness, the risk of making a VE decision on the basis of saving money could very well defeat the entire purpose of a hospital.

4.7 Noted deficiencies on day one

In every case the site preparation was wrong and violated code. In every case the start of the wall (base condition) was wrong. Had there been any challenge to the processes being employed on day one, or a pre-construction meeting to go over critical elements, or peer review, many of the more egregious defects may have been avoided. The AIA has no less than thirty-four documents that deal with project management and construction administration.[12]

5. Conclusion

5.1 Avoiding the pitfalls

5.1.1 The owner should define expectations up front and follow through

In the design phases both general contractors and major sub-contractors should be consulted regarding budgets and components. The owner should involve their facility managers and engineers in the design process from day one (programming) and should either make certain that the systems proposed do not exceed the skill level of their staff or, that training of their staff will be part of the construction process. The construction documents should be enough to ensure the quality of the building. As stated the perfect set of construction documents and the perfect contractor has yet to be seen. Periodic job site meetings with the owner in attendance should be mandatory. The contractor should provide and maintain a current schedule to enable the architects and engineers to review the critical elements of each phase. The owner should make it clear that problems are expected and they should not be glossed over but solved early.

Division Number	Division Title	Envelope Related	MEP and Controls Related	CS 1	CS2	CS3
1	General Conditions	Y	Y	√	√	√
2	Site	Y		√	√	√
3	Concrete	Y				√
4	Masonry	Y		√	√	√
5	Metals	Y				
6	Wood and Plastics	Y				
7	Thermal & Water Proofing	Y		√	√	√
8	Doors and Windows	Y		√	√	√
9	Finishes	Y		√		√
	Subtotal Envelope % non-conformance			66	55	78
10	Specialties	N				
11	Equipment	N				
12	furnishing	N				
13	Special construction	n/a				
14	Conveyance	n/a				
15	Mechanical		Y	√	√	√
16	Electrical		Y	√	√	√
	Subtotal Systems % non-conformance			100	100	100
	Total % non-conformance			72	63	82

Table 4. Identifies the topics covered within the CSI format that has been developed to cover all aspects of building construction. Y = yes, the Case Study included this section. √ = key elements within this division were found to be defective.

[12] AIA; G-Series.

5.1.2 The architect should produce a complete set of construction documents and a full service no matter what the delivery system[13]

Design/Build and Fast Track Projects have their place in the industry. However, even if budgets and contracts are made with preliminary designs or assumptions regarding a building's quality, the architect must not be swayed to reduce or eliminate services. In no case should he trust that the contractor (or some yet unidentified sub-contractor) will provide the detail that the architect would otherwise produce.

This is problematic for both parties. By either allowing or asking for "minimum" detail the contractor is telling the owner that he will provide industry standard and code compliant details and systems. He is, essentially, assuming the entire risk for the end product. Similarly the architect, by allowing the contractor that amount of leeway, may be faced with a claim of negligence. In a recent court decision an architect performing construction administration services could be held liable to third-party house guests injured by the contractor's failure to construct the project according to plans and specifications.[14]

5.1.3 The owner should consider peer review of the construction documents

Rarely done in the building industry peer review of the design (architectural and engineering) should be welcomed by the design professionals. Peer review would reduce the risk of having errors or omissions within the documents and, as with most peer reviews, challenge the designers to clarify and delineate their intent.

5.1.4 The owner should demand that only experienced professionals oversee the construction

Contractors can be intimidating to young professionals. In CS 3 the role of working with the contractor was delegated to a young designer. The project was sufficiently large enough that all of the elements that were found by the authors to be defective took weeks and months to construct thus allowing sufficient time for a seasoned professional to have the contractor replace defective work. Owners should insist that the architect provides experienced personnel during the construction phase of the work and that he, or she, is not replaced.

5.1.5 Thoroughly test layers of waterproofing systems

All three studies revealed leaks at the openings. Many specifications make reference to leak testing but it is often overlooked. The owner should demand that a formal process be included in the specifications and that it be performed by an independent party. The specification should require the leak testing be done before the exterior cladding is in place, IE after the weather-resistant barriers have been installed. The owner should have the testing company peer review all of the architect's details involving openings, base flashing, parapets, etc.

5.1.6 Inspect-inspect-inspect

In a project as critical as a hospital it is strongly advised to either expand the architect's visits to the site or hire an independent agent to fill in the gap between visits. As in peer

---

13 The traditional Design/Bid/Build is a delivery system. Others include Design/Build, Fast Track and various forms of Construction Management services.

14 Black + Vernooy Architects v. Smith, ___S.W.3d___, 2010 WL 5019659 (Tex.App. -- Austin 2010, no pet. h.),

review the third party involvement may result in the project being what the architect envisioned rather than having his and his owner's project discussed in front of a jury.

5.1.7 HVAC commissioning vs. test-adjust-balance

Construction documents always include a test, adjust and balance (TAB) section in Division 15 of the specifications. This specification require procedures usually performed by an independent agent to set-up air flows to design quantities, check and document the operation of components in the HVAC system such as, fans, compressors, pumps, etc. The data is then submitted to the engineer of record and the contractor with a list of deficiencies. In the case of the three case studies presented here, there was no follow-up. However, had they been follow-up, the root problems at these buildings would not have been solved. What a standard TAB does not do is integrate the various systems in the building with the HVAC system so that the whole building can be operated as a system.

The HVAC commissioning process provides a comprehensive overview of the building as a system. The procedures, methods, and documentation requirements in the document ASHRAE Guideline 1, HVAC Commissioning Process cover each phase of the commissioning process for all types and sizes of HVAC systems, from pre-design through final acceptance and post-occupancy, including changes in building and occupancy requirements after initial occupancy. Commissioning in accordance with ASHRAE Guideline 1 provides assurances and validates that the building's HVAC systems will perform as intended and will work with other systems in the building system as a whole. Commissioning procedures includes TAB.

Obviously commissioning is the preferred method. Although it cost more than TAB, it provides solutions to problems discovered during the construction process or immediately after occupancy. If the project team has an unsophisticated owner, builder, and/or design professionals VE may look attractive. Sophisticated owners and operators will probably recognize the value of commissioning and VE suggestions will be analyzed for what they really are. Regardless, it is the design teams responsibility to keep it on track.

5.1.8 Avoid being swayed by the latest and greatest "thing"

When we talk about sustainability or sustainable can we change that to "green". Green is not defined but it is understood by most. Sustainability/sustainable is also not defined but it is about to be defined by Department of Homeland Security in that when they talk about sustainability it will mean whether or not a building can sustain itself and continue to function to its design intent when it has suffered a disaster, natural or man-made.

Advocates in the theoretical "green world" get so preoccupied with finding and using "green materials" that they forget that at the end of the day the assembly still has to work as a high performance building, and that performance begins with the envelope. People are so enthralled with the idea of a "zero carbon footprint" that they are ignoring some of the basic principles of design that are already in our existing building codes and standards. This oversight, on their part, has resulted in the false understanding that existing codes and standards are not capable of dealing with the problems of energy conservation and they totally ignore that "organic architecture" has dealt with the problems they believe are new. Additionally the un-intended consequences of some of the proposed approaches to sustainability have yet to surface. What has surfaced, in 2010, is a class action lawsuit filed against the U.S. Green Building Council filed in U.S. District Court, Southern District of New York.

6. References

American Institute Of Architects (AIA) General Conditions of the Contract for Construction Standard Form of Agreement A201, 1997.
AIA Documents G711, Architects Field Report,1972.
AIA Documents G712, Shop Drawing And Sample Record, 1972.
AIA Documents G714, Construction Change Directive, 2001.
AIA Documents G716,Request For Information, 2004.
American Society of Heating Refrigeration and Air Conditioning Engineers (ASHRAE).
ASHRAE Applications Handbook, Chapter 7, Health Care Facilities, 2007.
ASHRAE Special Projects Committee SP-91, 2003 HVAC Design Manual for Hospitals and Clinics.
Cooley, J.D., Wong, W. C., Jumper, C. A., & Straus, D. C. (1988). Correlation between the prevalence of certain fungi and sick building syndrome. Occup. Environ. Med. 55, 579-584.
Harriman, L., Brundrett, G., Kittler, R. (2001). Humidity Control Design Guide for Commercial and Institutional Buildings.
Purdy, C.W., Clark, R.N., & Straus, D.C. (2009)Journal of Dairy Science, 92:6033-645 doi: 10.3168/jds.2009-2498. Ambient and indoor particulate aerosols generated by dairies in the southern High Plains.
Guidelines for Environmental Infection Control in Health-Care Facilities, Recommendations of CDC and the Healthcare Infection Control Practices Advisory Committee (HICPAC), U.S. Department of Health and Human Services, Center for Disease Control and prevention (CDC), 2003.

6

Air Quality Degradation: Can Economics Help in Measuring its Welfare Effects? A Review of Economic Valuation Studies

Eva Kougea and Phoebe Koundouri
Athens University of Economics and Business (AUEB)
Greece

1. Introduction

Air quality affects human's well being in various ways. Air providing the most important life-support function sustains human health and enables the existence of all ecosystems. Although clean air is considered to be a basic requirement for human health and well-being, economic development and population growth has resulted in a considerable deterioration of air quality. Human activities like the intensification of agriculture, industrialization, increasing energy use, the burning of fossil fuels and the increase in transportation have resulted to a rising cocktail of poisonous pollutants which impose many adverse effects on environment as a whole, our human health and life expectancy, ecosystems services, biodiversity, agricultural crops and building structures.

Under an economic perception excessive air pollution is partly caused by the existence of externalities mainly associated with missing markets or missing property rights. Social cost of air pollution is not covered by the polluting activities nor is reflected on market prices but it is imposed on everybody. The first part of this chapter refers to the theoretical framework of welfare economics and provides a brief description of fundamental issues arising in such environmental problems.

Economists have developed various approaches in order to be able to quantify values associated with non marketed environmental goods, in this case air quality. The next part of this chapter describes the currently available methods and techniques for estimating monetary values of changes in well-being due to air pollution. The methods will be presented can be classified into two broad categories, namely revealed preference techniques (cost of illness, human capital surveys, hedonic pricing, Quality Adjusted Life Years) and stated preference techniques (contingent valuation method and choice experiment).

The rest of the chapter provides a review of the existing literature on economic valuation studies which have attempted to elicit monetary values associated with air pollution, focusing in particular on human health, ecosystems services and buildings. Our aim is to provide a critical overview of the estimated benefits that public derive from an increase in air quality (or a reduction in environmental risk) by presenting a synthesis of the available quantitative empirical results. Finally, it follows a discussion regarding the usefulness of valuation studies in a policy context.

2. Theoretical concepts

The aim of this section is to provide a short introduction on the underlying economic theory with respect to non-marketed environmental goods such as air quality. Also explains the nature and the components of values individuals may derive from air quality improvement.

2.1 Air pollution as a public bad

Air quality under an economic perspective is a pure public good as it is characterized by both non-rivalry and non excludability in consumption. Respectively, air pollution can be seen as a public bad. Everyone breaths the air, it is not feasible for individuals to prevent themselves from exposure to ambient air pollution. In other words individuals are not free to vary independently the levels of the services of the environmental good they consumed, in this case air. This public good feature leads to market failure mainly associated with missing property rights and presence of externalities.

Air pollution is the most common form of an environmental externality. An externality arises when social or economic activities of one group of persons have an impact on another and when that impact is not fully informed or compensated for by the first group (Externe, 2005). Most economic activities emit a mixture of pollutants. These pollutants may create a health risk or disturb ecosystems. Though these damages are a cost to society, the polluter has no incentive to account for these damages when making decisions. Not only production of goods can generate external effects. In many cases consumption of goods can influence the size of externalities considerably (i.e. choice of mean of transport or fuel type). In both cases environmental costs of goods are ignored resulting to a lower private cost and thereby excessive production or consumption of the good that generates the externality.

Generally, from a microeconomic view, prices due to the presence of externalities are not the correct signals therefore market fails to obtain optimal social outcome. This means that private optimum divergences from social optimal allocation of the resources as the cost of air pollution is not covered by the polluting activities nor is reflected on market prices but it is imposed on everybody. Economic instruments should be employed in order to internalize externalities into the decision process. In order to do so first it is required to measure the damages to society in monetary terms and explore how external costs can be charged to consumers and producers.

2.2 Total economic value

The concept of total economic value is an essential theoretical framework for the implementation of economic valuation exercises. Individuals' preferences play the most important role on determining the value of an environmental good. Air provides a diverse array of goods and services which can be translated to values that directly or indirectly human society undertakes. Preferences actually reflect utilities individuals expect to derive from the consumption of a resource, given the needs the wants and the wishes. Total Economic Value (TEV) of a resource is the whole class of values that have a basis in human preferences (Karousakis & Koundouri, 2006).

Individuals are expected to derive utility from the consumption of a good or service. This is called actual use value and is composed of *direct use value* (i.e. commercial and recreational) and *indirect use value* (i.e. amenity value). People may also derive utility from retaining an option to a good or service. Option value reflects the potential future direct and indirect use values. Non-use values reflect individual preferences that may exist for an environmental good even when no actual or future use is expected to be made. Examples of non-use values are the *existence values* derived from the demand to preserve the existence of resources unrelated to any use and the *bequest* and *altruistic* values. The latter involve individuals' willingness to pay for an environmental resource for others, i.e. children or future generations.

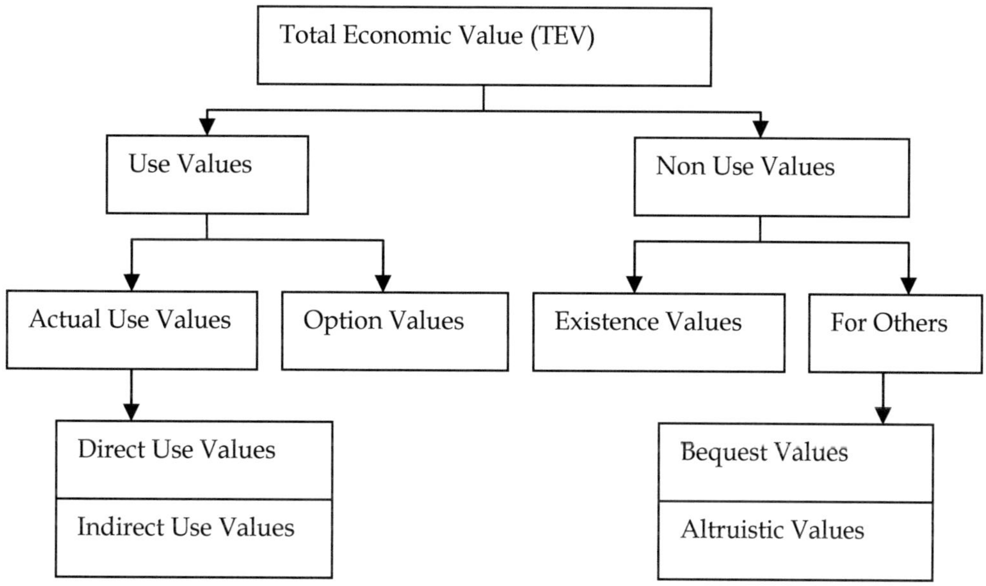

Fig. 1. Total Economic Value and its Components

Air pollution may influence all these elements of value. Economists in pollution control appraisal and in valuation exercises should consider all components of total economic value that are being influenced by the policy in question as TEV elements reflects the true value society undertakes from welfare changes due to an improvement or a deterioration of air quality.

3. Economic valuation techniques

The purpose of economic valuation is to capture the real economic values which contribute to the social welfare. Economic valuation refers to the assignment of money values, which have a particular and precise meaning, to non-marketed assets, goods and services (Pearce et al., 2002). A variety of techniques are available, the main methodological approaches of economic valuation can be broadly classified into revealed and stated preferences techniques.

Revealed Preferences (RP) techniques rely on market observations to capture the value of an environmental good that it is not itself traded in any market but is in a way connected with other marketed goods. From people's behavior in markets there is a possibility of isolating values of changes in environmental goods. Revealed preferences techniques include: Cost of Illness, Human Capital Surveys, Hedonic Pricing and the Quality Adjusted Life Year.

Hedonic Price Method (HPM) can be used to estimate the demand for air quality improvements. The most common implementation of hedonic pricing is in the housing markets. The theoretical structure behind hedonic pricing assumes that households, when making housing choices, consider a vector of characteristics including among others environmental factors as air pollution. Air pollution and environmental degradation can have direct impact on property values. It is expected that properties in areas with severe pollution problems will experience lower prices compared with properties located in areas of high air quality. HPM involves the estimation of a hedonic housing value equation with air pollution or other environmental factor as one of the housing attributes. Econometrical analysis will reveal each household's willingness to pay for a marginal change in air pollution from the hedonic housing value equation. A marginal WTP function for all households in the urban area examined is analogous to a demand curve for clean air.
Averting costs is the sum of the **defensive** and **preventive expenditures** that an individual is willing to make in order to avoid or reduce the risk of a negative impact. Rational individuals will be willing to spend money on activities that reduce the risk of air pollution up to the point where the marginal averting cost equals marginal value of reduced impact. The advantage of that technique is that makes direct use of market observable prices. With the same logic, in terms of human health, **Cost of illness** (COI) can be measured by the sum of the direct costs (medical costs, nursing care etc.) and indirect opportunity costs (i.e. loss of working days) associated with a disease and estimate the potential savings from the avoidance of the disease. **Human Capital Surveys** are employed to estimate the productivity loss measured in workdays due to illness. Also one can value loss of life based on foregone earnings associated with premature mortality. **Quality Adjusted Life Years (QALYs)** is an alternative measure to value reductions in health risks. Based on individuals' preferences it captures both the qualitative and quantitative aspects of life. An individual health is measured in terms of QALYs score. These scores are invariant across individuals in the same health state and range from 0 (death) to 1 (perfectly health), (Wagstaff, 1991). A similar measure closely related to QALYs is the **Disability Adjusted Life Years (DALYs)**. The main difference is that QALYs measure the value of a health profile relative to immediate health while DALYs incorporate a weighting factor that depends on age and measure the loss of longevity and health from an idealized health profile (Hammit, 2002).
Stated Preferences (SP) techniques are the most comprehensive and commonly used methods. In SP studies the market for the good is being constructed through the use of questionnaires. In this hypothetical market individuals have the opportunity to pay for an environmental improvement that will increase their utility or to accept compensation for an environmental deterioration that will decrease their utility. The most common forms of SP techniques are the Contingent Valuation Method and Choice Experiments.
Contingent Valuation (CV) is a survey based method technique in which a hypothetical market is being created and respondents are directly asked to express their willingness to pay contingent on some hypothetical change in the future state of environmental conditions (Mitchell & Carson, 1989). Alternatively, in case of an environmental deterioration this can be expressed as the minimum monetary compensation they would accept to tolerate the reduction in environmental quality. Individual's WTP estimates are very sensitive on the description of the contingent market, thereby information respondents have about the environmental good is determining the quality of the estimates. Thus the description of the environmental change in question and the contingent market must be presented carefully

with great detail. Of the same importance is the choice of a payment vehicle, i.e. increase in taxes vs. voluntary donations, as well as the form of the contingent valuation question i.e. an open-ended format vs. a double-bounded dichotomous choice mechanism. The literature on the contingent valuation method's advantages and disadvantages is large. There are many problems associated with CVM that may bias the value estimates (e.g. interviewing bias, non-response bias, strategic bias, embedding effects, yea-saying bias, hypothetical bias, information bias) (Bateman et al., 2002). NOOA Panel set the guidelines for the implementation of reliable contingent valuation exercises (Arrow et al., 1993). Recommendations among others refer to the choice formats i.e. dichotomous choice over other alternatives and to the choice of data collection, i.e. face to face interviews are preferable to web surveys.

Choice Experiment (CE) method is another survey based technique which can estimate the total economic value of an environmental good and its attributes as well as the value of more complex changes in several attributes. Each respondent is presented with a series of alternatives of an environmental stock, flow or service with varying levels of its price and attributes and asked to choose the most preferred option in each set of alternatives. A status quo alternative is included in each choice set. The selected attributes in each case are ought to be part of people's preference for the environmental change being considered and can be impacted by a management policy option (Bateman et al., 2002). The main advantage of this approach is the ability to measure the trade-off individuals make between different attributes and levels and that is possible to measure all elements of economic values like non-use values, i.e. existence and bequest. Another advantage of choice experiments is that eliminates or minimises several of the problems attached to contingent valuation like strategic bias, yea-saying bias, embedding effects (Bateman et al., 2002).

Due to the high cost and time demands needed to perform an original valuation study **Benefit Transfers** techniques have been developed. According to Rosenberg & Loomis (2001) benefit transfer is defined as the adaptation and use of existing economic information derived for specific sites under certain resources and policy conditions to new contexts or sites with similar resources and conditions. Navrud (2004) defines a typology of the most usual benefit transfer methods, namely the unit value transfer approach, unit transfers with adjustment to reflect site specific features and the benefit function transfer.

Finally it must be mentioned that a major part of the literature on economic valuation relies on constructing **dose-response function** and **damage functions.** They both reveal a technical or a biological relationship between quantities of a pollutant that affects a receptor with the physical impact on this receptor (Mitchell & Carson, 1989). These relationships can serve at the most as an aid in a monetary valuation study. Such dose-response functions are available for a range of pollutants and their physical impact on human health, building materials and agricultural crops.

Each technique described in this section has its own advantages and limitations. The selection of the appropriate technique should be case driven and be dependent of the policy examined. All available techniques differ in the data demands, assumptions and in the values that are able to capture. There are many methodological problems associated with the use of housing markets data or other revealed preference method to measure willingness to pay for clean air. The main limitation of these techniques is the capability to measure only a subset of the values people are willing to pay. They can capture only use values while they fail to isolate 'non-use' elements of value. Stated preference techniques have the advantage that can estimate non use values. Use values can be estimated by both revealed and stated preference techniques.

4. Economic assessment of air pollution: empirical evidence

In the subsequent section important applications of the valuation techniques that have been conducted to estimate social benefits associated with increased air quality are reviewed. The review is by no means exhaustive but aims to highlight the benefits of using economic valuation techniques as tools that can facilitate the formulation of stronger management policies that account more fully for the total economic value generated by air quality improvements. For quick reference, at the end of the chapter will be provided a table summarizing each study's mains features i.e. author, case study country, valuation technique and valuation result.

4.1 Air pollution human health and well-being

There is an increasing recognition that air pollution is a major environmental risk to health. Current concentrations of air pollutants have been associated with adverse long and short term health effects including an increase in mortality (Katsouyanni, 2003). According to the European Commission's Clean Air for Europe poor air quality results in several hundreds of thousands premature deaths in Europe each year, increased hospital admissions, extra medication and millions of lost working days (CAFE, 2005). In line with this, a large number of valuation studies trying to quantify environmental impacts on human health in monetary terms and elicit preferences for health and environmental policies that reduce the risk of illness or even mortality have appeared in the literature worldwide.

Many authors have reviewed studies addressing issues of economic valuation of air pollution and human health. Bell et al. (2008) review the literature on valuation studies assessing health consequences from greenhouses gases. The review was restricted to health benefits from air pollution exposure. Pearce (1996) also provides a summary of the main studies conducted to that day valuing health damages from air pollution focusing in developing countries. Preliminary work indicates that some air pollutants, like particulate pollution and lead, are associated with high damage in monetary terms for the developing world. Another review of the economic literature on the effects of environmental changes on public health, in both developed and developing world is presented by Remoundou & Koundouri (2009).

Contingent valuation studies dominate the literature addressing air pollution control assessment in both developed and developing world. In these studies health consequences of alternative air pollution mitigation strategies are presented in a valuation scenario and individuals are asked to state their maximum willingness to pay for the implementation of the policy under evaluation.

In Europe, mortality risk reductions issues expressed as extension in life expectancy are addressed by Desaigues et al. (2004) and Chilton et al. (2004). Desaigues et al. (2004) performed a contingent valuation study in France employing a questionnaire originally developed by Krupnick et al. (2002) in order to impute a value on air pollution mortality. Results from the survey used to provide estimates of value of statistical life and of value of life years. Each respondent expressed his/her willingness to pay annually during the next 10 years for a medical treatment that would reduce the risk of dying by 1 in 1000 and by 5 in 1000, estimated WTP is 412€ and 563€ respectively. Authors also review the valuation literature of a gain in life expectancy due to air pollution reductions. Another similar study is made by Chilton et al. (2004) who conduct a contingent valuation exercise in UK. In total 665 persons divided in 3 sub-samples were asked to value a bundle of health impacts of air pollution on months of extra life expectancy in normal health, months of extra life in poor health, avoiding hospital admissions, and avoiding breathing discomfort. Each sample was provided with different information regarding time horizon, i.e. one month, three months and six months avoidance of health symptoms. Respondents are asked about their annual WTP for the rest of their life. The mean annual WTP for chronic mortality is £ 29.52 for the one month sample, £ 30.21 for the three months sample and £38.73 for the six month sample. The value of one life year is £27,630, £9,430 and £6,040 for the one, three and six months sample respectively.

Focusing on human morbidity from air pollution, Navrud (2001) conduct a contingent valuation study to assess the benefits from morbidity risk reduction in Norway, Europe. This study aimed to elicit willingness-to-pay estimates to avoid additional days of seven lights health symptoms (coughing, sinus congestion, throat congestion, eye irritation, headache, shortness of breath and acute bronchitis) and asthma. The survey was conducted with face to face interviews of 1009 individuals. Mean WTP for an environmental program that would result to health risk reduction, i.e. avoidance of one additional day of health symptoms ranged from 99 NOK for coughing to 267 NOK for shortness of breath. WTP per person to avoid 14 additional days per year of light health symptoms ranged from 267 NOK for coughing to 812 NOK for shortness of breath. Authors also compared values estimates from this study with estimate from other similar US studies. Results showed that respiratory symptom days and asthma attacks are valued lower in Norway than in the US.

Carlsson & Johansson-Stenman (2000) implement a contingent valuation method to measure the benefits associated with increased air quality in Sweden. The mean WTP for a 50% reduction of harmful substances where the respondents live and work was about 2000 SEK/year (160SEK/month). WTP was found to increase in income, wealth and education. Also it was larger for men, members of environmental organizations, people living in big cities and people who own their house or apartment. Results seem to conform in general to the estimates of a previous stated preferences study of Halvoren (1996) in Norway. Results from a contingent valuation indicated that the WTP for a 50% reduction in air pollution due to reduced emissions from traffic is about 1250 SEK / year and person.

Dziegielewska & Mendelsohn (2005) employed a contingent valuation study to estimate the willingness to pay of Polish people to harmonize national air pollution standards with EU standards. Analysis designed to estimate a value of a 50% reduction in air pollution that in general correspond to EU standards for PM10 and a value of 25% reduction in air pollution. In total eight damages components of air pollution are considered namely mortality, incidence of bronchitis, asthma, minor health symptoms, visibility loss, material damages, damages to historical buildings and monuments and ecosystems damages. The payment vehicle used was a one-time increase in taxes. Data collection based on face to face interviews with 1,055 people. Results indicated that the total value as a fraction of GDP per capita in Poland was 0.77% for a 25% pollution reduction and 0.96% for a 50% reduction.

In the developing world a large number of contingent valuation studies addressing air pollution and either averted mortality or averted morbidity due to air pollution mitigation strategies have been conducted. In Taiwan these issues are addressed through contingent valuation by Alberini et al. (1997). The study of Alberini et al. (1997) demonstrates a contingent valuation exercise in three cities of Taiwan (Taipei, Kaohsiung and Hualien). In this study, people instead of being asked to state their WTP for a specified type of illness already defined for them, are asked to value their own illness experience and report their WTP to avoid it. The main advantage of that valuation design is that people are familiar with the commodity to be valued. Estimated WTP (in 1992 US$) to avoid an entire episode of cold is $20.45 for 1-day episode and $34.62 for a 5-day episode. If the episode is not cold

WTP is $30.73 and $52.01 respectively. Authors concluded that willingness to pay to avoid a day of illness is higher for a 1-day episode than for each day of a 5-day episode. Also the nature of the illness (i.e. cold or other) significantly affects WTP values. Another example of a contingent valuation in Taiwan can be found by Alberini & Krupnick (2000) who estimate willingness to pay to avoid minor respiratory illnesses associated with air pollution. Chestnut et al. (1997) investigate the benefits to human health due to reductions in particulate air pollution. In their study can be found summarized the results from a number economic valuation studies conducted in Bangkok. In this study WTP estimates from Bangkok are compared with US estimates in order to examine the transferability of air pollution control health benefits from the United States to developing countries. Authors concluded that Bangkok residents are willing to pay a higher share of their income to protect their health.

Mortality and morbidity effects of air pollution have been extensively studied through contingent valuation in China (Hammit & Zhou, 2006; Wang H. & Mullahy, 2006; and Wang Y. & Zhang, 2009). Hammit & Zhou (2006) employed a contingent valuation to estimate the economic value of air pollution related health risks in three diverse locations in China. This study is recognized as the first organized and peer-reviewed CV study on health effects in China. Authors valued colds, bronchitis and mortality related to air pollution. Sample's average median WTP to prevent an episode of cold ranges between US$3 and US$6, WTP to prevent a statistical case of chronic bronchitis ranges between US$500 and US$1000 and the value of statistical life ranged between US$4000 and US$17000.

Wang Y. & Zhang (2009) implement a contingent valuation exercise to quantify individuals' willingness to pay for improved air quality in the city of Ji'nan, China. A stratified sample of 1500 residents was chosen and WTP was elicited by conducting face to face interviews employing a series of hypothetical, open ended scenario questions. Almost 60% of respondents expressed a positive willingness to pay, WTP estimated at 100 Chinese Yuan per person, per year. Debriefing questions revealed that most respondents regard air quality improvement as a government responsibility, 40% of the respondents had no incentive to bear the costs of attempting better air quality, indicating a relatively low environmental consciousness.

Wang H. & Mullahy (2006) employed a contingent valuation method in an attempt to estimate the value of statistical life, i.e. the willingness to pay for reducing fatal risk by improving air quality in Chongqing, China. A sample of 500 respondents selected based on multistage sampling methods. Face to face interviews and a series of hypothetical open ended scenarios are employed to elicit WTP for air pollution reductions. WTP to save one statistical life estimated to be $34,458 while the mean annual income is $490. Interesting finding of this study is that people in China, unlike other developed countries, appear to consider clean air a luxury good.

Choice experiment currently offers the most promising approach to value non-marketed environmental goods. Even though this technique has been widely applied for the valuation of other environmental resources and services, the use of choice experiment on air pollution valuation literature is still limited. Rodriguez & Leon (2004) performed a choice experiment in Las Palmas Grand Canaria, Spain focusing on health effects caused by emissions from a large power plant. The policy measure to be valued involves the installation of filters that reduce emissions of particulates and other gases that cause adverse health effects. Installation would reduce the probability of becoming ill or suffering some episode of respiratory illness. A sample of 350 persons was randomly selected from the population and data collected through face to face interviews. Results showed that policy proposal would benefit people in the polluted suburb. Paternalistic altruism appeared to be present and significant for the large majority of the sample.

Banfi et al. (2007) attempted to estimate the benefits of an increase in local environmental quality in the Swiss cities of Zurich and Lugano by conducting a web-based choice experiment. In particular authors examine the impact of air pollution externalities on the rent for dwellings. Dataset comprise a representative sample of 394 households for Zurich and 241 households for Lugano. In each choice set respondents were asked to choose between three alternatives scenarios with varying environmental characteristics including air quality and monthly rent. The last alternative represented current dwelling situation. Estimation results showed that levels of air pollution are important factor when choosing a dwelling. People showed a positive and significant WTP for an improvement of environmental quality in the both urban areas. Good air quality along with low traffic noise exposure, were the highest valued attributes.

A choice experiment is also applied by Yoo et al. (2008) in an attempt to quantify the environmental costs of four air pollution impacts (mortality, morbidity, soiling damage, and poor visibility) in Seoul, Korea. Authors consider the trade-offs between price and other attributes of air pollution impacts and derive the marginal willingness to pay (WTP) estimate for each attribute. According to the results, the households' monthly WTP for a 10% reduction in the concentrations of major pollutants in Seoul was found to be approximately 5494 Korean won and the total annual WTP for the entire population of Seoul was about 203.4 billion Korean won.

The determinant role of economic assessment in accounting external damage into policy design process or in project appraisal has already been mentioned. In the literature, economic valuation techniques have also been used to provide economics ground for supporting investments in air pollution abatement enabling the application of a cost benefit analysis (Aunan et al. 1998; Larson, 1999, Li et al., 2004; Miraglia, 2007).

Aunan et al. (1998) assessed the cost and benefits of the implementation of a specific energy saving program in Hungary. Authors considered the possible reduced damage to public health, building materials and agricultural crops that may be obtained from reducing emissions of important air pollutants. Also accounted how the program contributes to reduced emissions of greenhouse gases. Analysis indicated that the main benefits from reducing the pollutants' concentrations relate to public health. The estimated benefits emerged from improved health conditions alone appear to exceed the investment needed to implement the program. Another study that has focused on the energy emission has been conducted by Larson et al. (1999) who employed a cost benefit analysis to assess the efficiency of five projects that would lead to 25% reduction in mortality risk due to reductions in particulate emissions in Russia. Economic assessment of the projects revealed a positive net present benefit of $40 million which justify the undertaking of the projects on economic ground. Miraglia, (2007) also performed a cost benefit analysis to assess the health, environmental and economic costs of the use of stabilized/ethanol mixture in the city of Sao Paolo. The quantitative environmental analysis resulted in a positive balance of US$ 2.851 billion. Finally in the city of Shanghai in China Li et al. (2004) performed an illustrative cost benefit analysis to assess air pollution controls. Health benefits of pollution reduction are compared with investments costs of the new technology strategies in question. Benefit- cost ratio revealed that are considerable net benefits to be derived from the implementation of new technology in China.

Sometimes in valuation exercises two different methods can be used. This gives the opportunity to researchers to compare the results obtained and test for the reliability of the valuation. Belhaz (2003) employed contingent valuation and hedonic price methods to estimate the benefits of clean air in Morocco from a 50% reduction of air pollution caused by road traffic in Rabat-Sale. In the valuation literature, hedonic studies have been applied to estimate a relationship between housing prices and other attributes, including health risks associated with air pollution. The value people place on reduced health risks through improved air quality can be inferred by their WTP for houses with better air quality. The first application of hedonic pricing to demonstrate the effect of air pollution to the housing prices was conducted by Ridker & Henning (1967) in USA. Results of that study indicated that air quality had a significant impact on housing prices. In particular it was estimated that reducing the sulphate level on 0.25 mg/day increased the value of houses between $84 and $245 (in 1960). From that day many studies have used a hedonic approach to estimate the relationship between houses' prices and air pollution. Komarova (2009) aims to calculate implicit prices of the environmental level of air quality in the city of Moscow on the basis of housing property rights. Database used contained approximately 20 thousand apartments. Explanatory variables included physical characteristics of the houses, environmental, neighborhood, socio-demographic and geographic data. Hedonic analysis showed that ecological variable had a negative price while increasing the level of air contamination from carbon monoxide, nitrogen dioxide, sulphur dioxide and particles. Brucato et al (1990) and Won Kim et al. (2003) also applied hedonic pricing to isolate and measure the impact of air pollution. Smith & Huang (1995) and Delucchi et al. (2002) performed a meta-analysis of cross sectional hedonic studies. Results indicate that hedonic price does not capture total health costs of air pollution because individuals are not fully informed about all of the health effects. Additionally the value of improved air quality is underestimated as this technique does not capture non use elements of value.

Other techniques have also being applied in the valuation literature of air pollution and human health. Hubbel (2006) implemented QALYs in the analysis of air pollution regulations. Precisely he explored the implications of the QALY approach to measure impacts of air pollution regulations, in particular US EPA Heavy Duty Engine/Diesel Fuel, and he investigated the potential use of QALY in cost-benefit analysis. Damigos (2005) used a damage cost approach to examine the implementation of a process that would reduce NOx emissions that are produced by the operation of an oil steam turbine power plant. Two alternatives are considered: a baseline do-nothing scenario, corresponding to 7332 tn/year emissions of NOx, and a NOx emissions reduction scenario corresponding to 2985 tn/year. The implementation of NOx reduction measures will result in an environmental improvement that is calculated at 3,350,000 euro per annum. Vrhovcak et al. (2005) used a damage cost method to estimate the damages to human health resulting from electricity production in Croatian thermal power plants. The following health end-points, were considered: acute mortality ($148,500), chronic mortality ($1,375,000), hospital treatment of cardiovascular diseases ($10,300), restricted activity days ($100), hospital treatment of respiratory diseases ($10,300), chronic bronchitis in children ($300) in adults ($138,000) and congestive heart failures in elder individuals ($10,300). In the brackets can be found the estimated total external costs i.e. the value of damage in 1995 US$ from nine Croatian fossil fuel-fired power plants. Cesar et al. (2002) calculated the health benefits from a number of pollution reduction scenarios using appropriate dose response functions from the literature.

Mexico City suffered from high levels of ozone and particulate pollution for years. Authors valued the avoided cases of morbidity and premature mortality by employing cost of illness and forgone earnings, willingness for avoided morbidity and willingness to pay for avoided mortality. Results suggest that a 10% reduction in ozone was about $760 million (in 1999 U.S. dollars) annually and the benefit for a 20% reduction in ozone and PM was about $1.49 billion annually. In India, Gupta (2006) employed a cost of illness approach to estimate the monetary benefits to individuals from health damages avoidance due to air pollution reductions. Health costs due to adverse effects of air pollution on health were considered as the sum of the loss in wages due to workdays lost and the expenditures made by households on mitigating activities. Murty et al. (2003) used household data to analyze the impact of higher levels of Suspended Particulate Matter (SPM) in the Indian metropolitan cities of Delhi and Kolkata. Health production function and demand functions for averting and mitigating activities were estimated. Study revealed that the annual marginal benefits to a typical household ranged from Rs 2086 in Delhi to Rs 950 in Kolkata if the level of SPM is reduced from current average level to the prescribed safe level.

Last, a body of research in economics employing self-reported happiness, or life satisfaction approach to examine the benefits that could be gained from higher levels of a non-marketed environmental good. Focusing in particular on air quality Welsch (2006) explores the relationship between air pollution and prosperity in ten European employing self reporting well being and it examines how well being varies with air quality and prosperity. The rate at which they trade off against each other is being calculated by measuring marginal disutility of pollution, as well as the marginal utility of income. It was found that a decrease in the NO_2 concentrations is valued at $398 per person per year in Greece up to $1858 in Luxembourg (in 1997). Differences between values reflect the fact that people place more value on environmental improvement in richer than in poorer countries. In relative terms, the value of reduced NO_2 ranges from 2.3% of per capita income (UK) up to 9.8% (Spain). With respect to lead concentrations, the estimated values ranging from $657 (Greece) up to $3113 (Denmark). The total value of air quality improvement (NO_2 and lead) ranges from $1076 (Greece) up to $3859 (Denmark) while the total value as a percentage of income ranges from 5.1% (Luxembourg) up to 18.2% (Denmark).

All studies mentioned have focused on ambient air quality. Very few valuation studies have investigated the benefits generated by improved indoor air quality. Chau et al. (2007) combined revealed and stated preferences techniques to estimate the monetary benefits gains from improved indoor air quality. It was found that significant gains are derived from an increase in indoor air quality.

4.2 Air pollution ecosystems, buildings materials and cultural heritage

It is well known that air pollution affects ecosystem's good and services. Many studies have attempted to impute a monetary value to these damages. MacMillan (2001) in a report prepared for DEFRA provides a review on the literature on ecosystem valuation in the context of air pollution and describes the main challenges confronting valuing ecosystems recovery. Author concluded that studies from North America and Scandinavia dominate the international literature on valuing ecosystems effects of air pollution. Most of these studies are concerned with impacts of air pollution on forest growth and timber. Concluding remark is that assessing the benefits of ecosystem recovery from air pollution is not a straightforward procedure. The main reason for this is the great uncertainty surrounding both future environmental damage of air pollution and future environmental recovery from air pollution reductions. Scientific understanding of the complex ecosystem function in many cases is very limited.

Contingent valuation can incorporate such issues of uncertainty. An example of a contingent valuation implementation in the ecosystem valuation literature is provided by Macmillan et al. (1996) who estimate willingness to pay of the Scottish population for uncertain recovery and damage scenarios from reduced acid rain deposition in the semi-natural uplands of Scotland. In order to deal with uncertainty, it has been used a split-sample survey format presenting six alternative ecosystem recovery levels and damage levels scenarios. Approximately 1000 households were sampled by mail. Average household WTP, elicited using a dichotomous choice format, for abatement of acid rain ranged from £247 to £351 depending on the scenario. WTP was statistical significantly influenced by the level of future damage but not by future recovery level.

Pollution and climatic parameters may also have a direct effect on several materials. Rabl (1999) value the damage to building materials linked to air pollution in France. Analysis provides estimates for the damage cost of air pollution on historical buildings focusing on the effects of acidic deposition on corrosion and deposition of soiling. Aunan et al. (1998) in a project assessment accounted for the effects of air pollution on buildings and indicated that there are significant benefits to be gained due to reduced replacement and maintenance costs for buildings materials (US$ 30-35 million annually in Budapest only). Regarding cultural heritage, there can be found very few valuation studies undertaken. Yet historical heritage damages, as well as ecosystems damages, are important and the literature underestimates total damages by omitting them. Results from a Polish case study of Dziegielewska & Mendelsohn (2005) showed that if these components were not valued, the results of that study would be underestimated by 13-16%. Most of the studies tackling these issues are mainly employing a contingent valuation method (Pollicino & Maddison, 2002; Morey et al., 2002; Navrud & Strand, 2002). Pollicino & Maddison (2002) performed a contingent valuation method to value aesthetic changes in the Lincoln Cathedral, UK due to air pollution, Morey et al. (2002) employed a contingent valuation to value acid deposition injuries to marble monuments in Washingtom, DC, and Navrud & Strand (2002) employed a contingent valuation exercise to value the damages from air pollution on the Nidaros Cathedral in Norway. Even limited this part of the literature ought to be helpful to agencies and organizations whose mission is the protection and preservation of cultural heritage, i.e. historic buildings, monuments etc., from exposure to pollution. These studies provide public values for cultural heritage goods and can provide much information regarding social perceptions.

5. The use of valuation results in policy design

The strongest argument for assigning economic values is the use of valuation results in policy design. Theory underlies that if external effects are omitted from appraisal there is high risk that non-marketed goods will be under-supplied while non-marketed bad will be oversupplied. As it was mentioned in the first section of this chapter since the market mechanism is not able to capture the total value of environmental goods, public intervention is needed in order externalities be internalized into the decision process. It is important economic decisions to be made compatible with social objectives, i.e. efficiency and equity considerations. Preferences elicitation for different socio-economic groups and knowledge of the marginal valuation each group attaches to environmental improvements through valuation studies allows for equity considerations to be taken into account in the formulation of policy design (Remoundou & Koundouri, 2009). Economic estimates give also information regarding the best use of available resources i.e. the option that has the lowest opportunity cost or the lowest value to be sacrificed. Monetary value assessment allows the ranking of alternative policy options through the implementation of cost-benefit analysis for policy guidance. There are many issues arising on how to introduce monetary valuation into public decision making. Pearce (2001) formulates some recommendations of getting CBA into the process of decision making. Monetary benefit estimates can serve many purposes. Bonnieux & Rainelli, (1999) summarized the usefulness of valuation exercises. First of all valuation exercise can contribute to public debate and awareness concerning specific (environmental) problems. Translate environmental changes into money values can serve as a readily understandable indicator of environmental damage or potential benefits. Secondly monetary benefit estimates can influence particular decisions by employing a cost benefit analysis (or any other way of comparing costs and benefits). So the use of such estimates can help policy makers to identify the optimal alternative among competing options and to support and justify decisions taken by government agencies (Bonnieux & Rainelli, 1999).

However limitations of the existing literature have been identified. The major obstacle that has restricted the use of benefits estimates is the uncertainty or the low credibility of the estimates. Responsible causes for this weakness can be: (Kuik et al. in Navrud, 1992)

1. Uncertainty about the most elementary physical dose-response relationship
2. Hypothetical character of the estimated money values.
3. Choice of the correct discount factor for discounting costs and benefits which occur on different points in time
4. Transboundary pollution problems and the choice of the relevant population affected in valuation exercise.

6. Conclusion

This review aimed to serve as a starting point for research on economic valuation estimates. It gives an elaborate cross-section of studies that are described in journals as well as studies financed by government agencies of which only reports are available. Findings of the literature indicate that society undertakes major benefits from reduction of air pollution level. Internationally the use of monetary valuation techniques in project appraisal has increased significantly. Techniques such as the contingent valuation which can estimate values for both users and non-users are being intensively used and are being constantly refined and extended. Findings of the literature review can be summarized in the followings:

- Studies of the developing world overweight number of studies conducted in developed countries.
- Limited application of choice experiment in health valuation studies related to air pollution.
- Limited number of studies assessing indoor air quality impacts on human health.
- Studies showed that are significant gains estimated to be derived from air pollution control instruments.
- Multidisciplinary research and close collaboration between economists and other scientists must be promoted in order more credible economic estimates to be produced.

- The use of hedonic pricing underestimates values generated by improved air quality as it fails to capture non use elements of value.
- Improve scientific information is needed in order to be able to produce more reliable dose - response functions and thus more credible economic estimates of the values.
- Effects of air pollution on ecosystems and cultural heritage appear to be a significant damage to the society. Omitting these components of damage would result to a backwards bias in the estimates.
- Recent years have seen a growing interest in the potential of producing generally applicable estimates for the valuation of non-marketed environmental goods to another policy site. Many studies have addressed these issues and have tested the validity of benefit transfers. Meta-analysis of the existing results is another common application.

It follows a table summarizing valuation studies mentioned in this chapter.

Authors	**Study Area**	**Valuation Technique**	**Environmental Impact / Good to be Valued**	**Results**
Carlson & Johansson - Stenman, (2000)	Sweden	Contingent Valuation	Benefits from improved air quality	WTP for a 50% reduction of harmful substances where the respondents live and work estimated at about 2000 SEK/year
Navrud, (2001)	Norway	Contingent Valuation	Air pollution and human morbidity: seven lights health symptoms and asthma.	Authors produced mean WTP estimates for the avoidance of 1 and 14 additional days 'light' health symptoms per year.
Halvoren B. (1996)	Norway	Contingent valuation	Health benefits from a decline in air pollution due to reduced emissions from traffic.	WTP for a 50% reduction in air pollution estimated at 1250 SEK/year and person.
Dziegielewska & Mendelsohn, (2005)	Poland	Contingent Valuation	Adverse impact of air pollution on Health: mortality, incidence of bronchitis, asthma, minor health symptoms; Visibility loss, Material damages to historical buildings and Ecosystems damages	Authors produced WTP estimates for a 50% reduction and a 25% reduction in air pollution. Also estimated damage of air pollution as a fraction of the GDP. Historical heritage and ecosystem damage appeared to be significant components of total damage

Authors	Study Area	Valuation Technique	Environmental Impact / Good to be Valued	Results
Desaigues et al. (2004)	France	Contingent Valuation	Air Pollution and Human	Annual WTP for the next 10 years for a medical treatment that would reduce the risk of dying by 1 in 1000 is 412€ and by 5 in 1000 is 563€.
Banfi et al., (2007)	Zurich and Lugano	Choice experiment	Impact of air pollution externalities on human welfare. Noise pollution and electro smog were also included into the analysis	Mean WTP for air quality improvements from bad to good is 198 in Zurich and 151 in Lugano (in CHF per month)
				Mean WTP for air quality improvements from bad to medium is 198 in Zurich and 94 in Lugano (in CHF per month)
				Mean WTP air quality improvements from medium to good is 70 in Zurich and 57 in Lugano (in CHF per month)
Rodriguez & Leon, (2004)	Grand Canaries, Spain	Choice Experiment	Health impacts of energy production	Results showed that people would benefit from the installation of filters that reduce particulate emissions. Significant and positive altruistic values
Yoo et al. (2008)	Seoul	Choice Experiment	Environmental costs of air pollution impacts (mortality, morbidity, soiling damages and poor visibility)	Households' monthly WTP for a 10% reduction in the concentrations of major pollutants was approximately 5494 Korean won. Total annual WTP for the entire population was about 203.4 billion Korean won

Authors	Study Area	Valuation Technique	Environmental Impact / Good to be Valued	Results
Wang H. & Mullahy (2006)	China	Contingent Valuation	Reduced fatal risk by improving air quality	Value of statistical life estimated to be $34,468
Wang Y. & Zhang, (2009)	China	Contingent Valuation	Effects of air pollution on human health	Mean WTP was estimated at 100 Chinese yan per person per year
Hammit & Zhou, (2006)	Three diverse locations in China	Contingent Valuation	Benefits from improved air quality	Median WTP to prevent an episode of cold ranges between US$3 and US$6, WTP to prevent a statistical case of chronic bronchitis ranges between US$500 and US$1000 and the value of statistical life ranged between US$4000 and US$17000.
Belhaz, (2003)	Morocco	Contingent Valuation and Hedonic Pricing	Air pollution road traffic Benefits from clean air	Authors produce WTP estimates for a 50% reduction of air pollution
Alberini et al., (1997)	Three cities of Taiwan	Contingent Valuation	Air Pollution – Human Health	Authors produce WTP to avoid the recurrence of an episode of minor respiratory illness identical to that most recently experienced by the respondent
Alberini & Krupnick, (2000)	Taiwan	Contingent Valuation	Air Pollution Human Health	Authors produced WTP estimates to avoid minor respiratory illnesses associated with air pollution.
Chestnut et al., (1997)	Bangkok	Contingent Valuation	Benefits to human health due to reductions in particulate pollution	Authors produce WTP estimates to avoid three minor respiratory health endpoints characterized by different severity: one-symptom day, one restricted activity day and one work-loss day.

Authors	Study Area	Valuation Technique	Environmental Impact / Good to be Valued	Results
Li et al., (2004)	Shnaghai	Cost Benefit Analysis	Health benefits from pollution reduction	Benefit cost ratio revealed considerable net benefits to be derived
Gupta, (2006)	India	Cost of illness	Monetary benefits to individuals from avoided health damages due to reductions in air pollution	Results indicated that the mean worker from Kanpur would gain Rs 165 per year if air pollution were reduced to a safe level
Welsch, (2006)	10 European Countries	Life satisfaction approach	Effect of air pollution on well- being	It was found that a decrease in the NO_2 concentrations is valued at $398 per person per year in Greece up to $1858 in Luxembourg (in 1997).
Rabl, (1999)	France	Damage Cost	Ambient Air Pollution and Historical Buildings	Analysis provides estimates for the damage cost of air pollution on historical buildings focusing on the effects of acidic deposition on corrosion and deposition of soiling.
Chilton et al., (2004)	UK	Contingent Valuation	Health benefits of improved air pollution in terms of mortality and morbidity	The mean annual WTP for chronic mortality is £ 29.52 for the one month sample, £ 30.21 for the three months sample and £38.73 for the six month sample.
Larson et al., (1999)	Russia	Value of a Statistical Life/Cost-Benefit Analysis	Particulate emissions and mortality	Economic assessment revealed a positive net benefit of $40million
Damigos, (2005)	Crete, Greece	Damage Costs	Air Pollution - energy	Environmental improvement due to reduced NOx emissions calculated at 3.350.000 € per year

Authors	Study Area	Valuation Technique	Environmental Impact / Good to be Valued	Results
Aunan et al., (1998)	Hungary	Cost-Benefit Analysis	Air Pollution and public health, buildings materials and agricultural crops	Estimated annual benefits of improved health conditions exceeds the investments needed
Miraglia, (2007)	Brazil	Cost-Benefit Analysis	Air Pollution	Quantitative environmental analysis resulted in a positive balance of 2851US$
Vrhovcak et al., (2005)	Croatia	Damage cost	External costs of electricity generation in terms of human health	Authors estimated external cost for a bundle of health endpoints
Cesar et al., (2002)	Mexico	Dose – response functions/ Cost of Illness	Health benefits from a number of pollution reduction scenarios	Results suggest that a 10% reduction in ozone was about $760 million (in 1999 U.S. dollars) annually and the benefit for a 20% reduction in ozone and PM was about 1.49 billion annually.
Murty et al. (2003)	Delhi and Kolkata	Health production functions and demand functions for averting and mitigating activities	Air pollution	Annual marginal benefits to a typical household ranged from Rs 2086 in Delhi to Rs 950 in Kolkata if the level of SPM is reduced from current average level to the prescribed safe level.
Komavora, (2009)	Moscow	Hedonic pricing	Relationship between air pollution and housing prices	Ecological variable had a negative sign while increasing the level of air contamination from carbon monoxide, nitrogen dioxide, sulphur dioxide and particles.
MacMillan, (1996)	Scotland	Contingent Valuation	Effect of reduced acid rain deposition in the ecosystems	Average WTP ranged from £247 to £351 depending on the scenario

Authors	Study Area	Valuation Technique	Environmental Impact / Good to be Valued	Results
Pollicino & Maddisson, (2002)	Lincoln Cathedral, UK	Contingent Valuation	Aesthetic changes on historical buildings due to air pollution	All studies produced WTP estimates for a policy option that would reduce harmful pollutants
Morey et al., (2002)	Washington	Contingent Valuation	Acid deposition injuries to marble monuments	
Navrud & Strand, (2002)	Nidaros Cathedral, Norway	Contingent Valuation	Air pollution and damages to buildings	

Table 1. Summary of the Valuation Studies

7. References

Alberini, A.; Cropper, M.; Fu, T.T.; Krupnick, A.; Liu, J.T.; Shaw, D. & Harrington, W. (1997). Valuing health effects of air pollution in Developing countries: the Taiwan experience. *Journal of Environmental Economics and Management,* Vol. 34, pp. 107-26

Alberini, A. & Krupnick, A. (2000). Cost of Illness and Willingness to Pay estimates of the Benefits of Improved Air Quality: Evidence from Taiwan. *Journal of Land Economics,* Vol. 76, No.1, pp. 37-53

Arrow, K.; Solow, R.; Portney, P.R.; Leamer, E.E & Radner, R.H. (1993). Report on the NOOA Panel on contingent valuations, natural resource damage assessment under the oil pollution act of 1990. *Federal Register, Vol.* 58 pp. 4601-4614

Aunan, K.; Patzay, G.; Aaheim, H.A. & Seip, H.M. (1998). Health and Environmental Benefits from Air Pollution Reductions in Hungary. *The Science of the Total Environment* Vol. 212, pp. 245-268

Banfi, S.; Filippini, M. & Horehajova, A. (2007). Using a Choice Experiment to Estimate the Benefits of a Reduction in Externalities in Urban Areas with Special Focus on Electrosmog. *Centre for Energy Policy and Economics (CEPE),* Working Paper No. 57, ETH (Swiss Federal Institute of Technology), Zurich

Bateman, I.; Carson, R.; Day, B.; Hanemann, M.; Hanley, N.; Hett, T.; Jones-Lee, M.; Loomes, G.; Mourato, S.; Ozdemiroglu, E.; Pearce, D.; Sugden, R. & Swanson, J. (2002) *Economic Valuation with Stated Preferences Techniques: A manual.* Cheltenham, Edward Elgar, ISBN:9781840649192

Belhaj, M. (2003). Estimating the benefits of clean air; Contingent valuation and hedonic price methods. *International Journal of Global Environmental Issues*, Vol. 3, No. 1, pp. 30-46

Bell, M.L.; Davis, D.L.; Cifuentes, L.A.; Krupnick, A.J; Morgenstern, R.D. & Thurston, G.D. (2008). Ancillary human health benefits of improved air quality resulting from climate change mitigation. *Journal of Environmental Health* Vol. 7, No. 41

Brucato, P.F.; Murdoch, J.C. & Thayer, M.A. (1990). Urban Air Quality Improvements: a Comparison of Aggregate Health and Welfare Benefits to Hedonic Price Differentials. *Journal of Environmental Management* Vol. 30, No. 3, pp. 265-279

Bonniex, F. & Rainelli, P. (1999). Contingent Valuation Methodology and the EU institutional framework. In Bateman, I.J., Willis, K.G. (eds.) *Valuing environmental preferences, Theory and practice of the contingent valuation method in the US, EU and Developing Countries.* New York, Oxford University Press pp. 585-612

CAFÉ (2005) Commission Staff Working Paper, Annex to: The Communication on Thematic Strategy on Air Pollution, and the Directive on Ambient Air Quality and Cleaner Air for Europe, SEC (2005) 1133

Carlsson, F. & Johansson-Stenman, O. (2000). Willingness to pay for improved air quality in Sweden. *Journal of Applied Economics* Vol. 32, pp. 661-669

Cesar, H.; Borja-Aburto, V.H.; Dorland, K.; Munoz Cruz, R.; Brander, L.; Cropper, M.; et al. (2002). Improving air quality in metropolitan Mexico City: an economic valuation, *Policy Research Working Paper Series 2785, The World Bank, Washington, DC*

Chau, C.K.; Hui, W.K. & Tse, M.S. (2007). Evaluation of health benefits for improving indoor air quality in workplace. *Environment International,* Vol. 33, No.2, pp. 186-198

Chestnut, L.G.; Ostro, B.D. & Vichit-Vadakan N. (1997). Transferability of Air Pollution Control Health Benefits Estimates from the United States to Developing countries: Evidence from the Bangkok study. *American Journal of Agricultural Economics,* Vol. 79, pp. 1630-1635

Chilton, S.; Covey, J.; Jones-Lee, M.; Loomes, G. & Metcalf, H. (2004). *Valuation of health benefits associated with reductions in air pollution, final report;* Department for Environment Food and Rural Affairs, London, UK.

EXTERNE (2005). *Externalities of Energy. Methodology 2005 Update* Bickel P. and Friedrich R. (eds), European Commission, EUR 21951 ISBN 92-79-00423-9

Damigos, D. (2005). Economic aspects and environmental cross-media effects in energy production. *WSEAS Int. Conference: Energy, Environment, Ecosystems, Sustainable Development,* Vouliagmeni Greece, IASME Transactions Vol. 3, No. 2, pp. 325-330

Delucchi, M.A.; Murphy, J.J. & McCubbin, D.R. (2002). The health and visibility costs of air pollution: a comparison of estimation methods. *Journal of Environmental Management* Vol. 64 pp. 139-152

Desaigues, B; Rabl, A; Ami, D.; My, K.B.; Masson, S.; Salomon, M.A.& Santoni, L. (2004). Monetary valuation of air pollution mortality: current Practice, Research Needs and Lessons from a Contingent valuation. *Universite de Strasbourg,* Strasbourg, Alsace, France

Dziegielewska, D. & Mendelsohn, R. (2005). Valuing air quality in Poland. *Journal of Environmental and Resources Economics,* Vol. 30, No.2, pp. 131-163

Gupta, U. (2006). Valuation of urban air pollution: a case study of Kanpur in India. SANDEE Working Paper No. 17-06, *South Asian Network for Development and Environmental Economics (SANDEE),* Kathmandu, Nepal

Halvoren, B. (1996). Ordering effects in contingent valuation surveys: Willingness to pay for reduced health damage from air pollution. *Journal of Environmental and Resources Economics,* Vol. 8, pp. 485-99

Hammit, J. K. (2002). QALYs vs WTP. *Journal of Risk Analysis,* Vol. 22, No.5, pp. 985-1001

Hammit, J.K. & Zhou, Y. (2006). The economic value of air pollution related health risks in China: a contingent valuation study. *Journal of Environmental and Resource Economics,* Vol. 33, No. 3, pp. 399-423

Hubbell, B.J. (2006). Implementing QALYs in the Analysis of Air Pollution Regulations. *Journal of Environmental and resources Economics,* Vol. 34, No. 3, pp. 365-384

Karousakis, K. & Koundouri, P. (2006). A Typology of Economic Instruments and Methods for Efficient Water Resources Management in Arid and Semi-Arid Regions, in *Water Management in Arid and Semi-Aid Regions: Interdisciplinary Perspectives.* Edward-Elgar Publishing, p. 161-187

Katsouyanni, K. (2003). Ambient air pollution and health. *British Medical Bulletin* Vol. 68, No. 1, pp. 143-156

Komavora, V. (2009). Valuing Envirnmental Impact of Air Pollution in Moscow with Hedonic Prices. *World Academy of Science, Engineering and Technology*, Vol. 7, pp. 319-326

Krupnick, A.; Alberini, A.; Cropper, M.; Simon, N.; O'Brien, B.; Ron Goeree, R. & Heintzelman, M. (2002). Age, Health and the Willingness to Pay for Mortality Risks reductions: A Contingent Valuation Survey of Ontario Residents. *Journal of Risk and Uncertainty* Vol. 24, No. 2, pp. 161-186

Kuik, O.; Navrud, S. & Pearce, D.W. (1992). Benefit estimation and environmental decision making. In Navrud S. (eds), *Pricing the European Environment,* Oslo, Scandinavian University Press

Larson, B.A.; Avaliani, S.; Golub, A.; Rosen, S.; Shaposhnikov, D.; Strukova, E.; Vincent, J.R. & Wolff, S.K. (1999). The economics of air pollution health risks in Russia: a case study of Volgograd. *World Development* Vol. 27, No.10, pp. 1803-1819

Li, J.; Guttikunda, S.K.; Carmichael, G.R; Streets, D.G; Chang, Y-S. & Fung, V. (2004). Quantifying the human health benefits of curbing air pollution in Shanghai. *Journal of Environmental Management* Vol. 70, No.1, pp. 49-62

Macmillan, D.; Hanley, N. & Buckland, S. (1996). A Contingent Valuation Study of Uncertain Environmental Gains. *Scottish Journal of Political Economy,* Vol. 43, No. 5, pp. 519-533.

MacMillan, D. (2001) Valuation of Air Pollution Effects on Ecosystems: A Scoping study. A report prepared to the *Department for Environment, Food and Rural Affairs. University of Aberdeen*

Miraglia, S.G.E.K. (2007). Health, environmental and economic costs from the use of a stabilized diesel/ethanol mixture in the city of Sao Paolo. *Cadernos de saude publica* Vol. 23, pp. 559-569

Mitchell, R.C & Carson, R. (1989). *Using Surveys to Value Public Goods: The Contingent Valuation Method.* Resources for the Future, Washington, D.C.

Morey, E.R.; Rossmann, K.G.; Chestnut, L.G. & Ragland S. (2002). Valuing Reduced Acid Deposition Injuries to Cultural Resources: Marble Monuments in Washington, D.C. In *Valuing Cultural Heritage: applying environmental techniques to historic buildings, monuments and artifacts,,* Navrud, S. & Ready, R.C. (eds.) Cheltenham: Edward Elgar Publishing Limited, UK

Murty, M.N.; Gulati, S.C. & Banerjee A. (2003). *Health Benefits from Urban Air Pollution Abatement in the Indian Subcontinent.* E/236/2003, Delhi: Institute of Economic Growth

Navrud, S. (2001).Valuing health impacts from air pollution in Europe. *Journal of Environmental and Resources Economics,* Vol. 20, No.4, pp. 305-329

Navrud, S. & Strand, J. (2002). Social Costs and Benefits of Preserving and Restorig Nidaros Cathedral. In *Valuing Cultural Heritage: applying environmental techniques to historic buildings, monuments and artifacts,* Navrud, S. and Ready, R.C. (eds) pp. 31-39, Chentlenham: Edward Elgar Publishing Limited, UK, ISBN 1 84064 079 0

Navrud, S. (2004). Value transfer and environmental policy. In *The International Yearbook of Environmental and Resources Economics 2004/2005: a survey of current issues* Tietenberg, T and Folmer H. (eds.) Chapter 5, pp. 189-217 Edward Elgar Publishing, Cheltenham, UK and Northampton, MA, USA.

Pearce, D. (1996). Economic Valuation and Health Damage from Air Pollution in the Developing World. *Energy Policy* Vol. 24, No. 7, pp. 627-630.

Pearce D. (2001) Annex II: Integrating cost-benefit analysis into the policy process in Howarth, A., Pearce D.W., Ozdemiroglu, E., Seccombe-Hett, T., Wieringa, K.,

Streefkerk, C.M. de Hollander, A.E.M, RIVM report 481 505 024 *Valuing the benefits of environmental policy:* The Netherlands, Bilthhover, RIVM.

Pearce D.W. and Ozdemiroglu, E et al. (2002) *Economic Valuation with Stated Preference Techniques. Summary Guide.* Department for Transport, Local Government and the Regions: London, UK

Pollicino, M. & Madisson, D. (2002). Valuing the Impacts of Air Pollution on Lincoln Cathedral. In *Valuing Cultural Heritage: applying environmental techniques to historic buildings, monuments and artifacts.* Navrud, S. and Ready, R.C. (eds) Chentlenham: Edward Elgar Publishing Limited

Rabl, A. (1999). Air pollution and Buildings: an Estimation of Damage Costs in France. *Environmental Impact Assessment Review*, Vol. 19, No. 4, pp. 361-385

Remoundou, K. & Koundouri P. (2009). Environmental effects on public health: An economic perspective. *International Journal of Environmental Research and Public Health,* Vol.6, No.8, pp. 2160-2178

Ridker, R.G. & Henning J. A. (1967). The determinants of residential property values with special reference to air pollution. *The Review of Economic and Statistics,* Vol. 49, No. 2, pp. 246-257

Smith, V.K & Huang, J-C. (1995) Can Markets Value Air Quality? A Meta-Analysis of Hedonic Property Value Models. *The Journal of Political Economy* Vol. 103, No.1, pp. 209-227

Rodriguez, V. M. & Leon, C. (2004). Altruism and the Economic Values of Environmental and Social Policies. *Journal of Environmental and Resources Economics,* Vol. 28, No.2, pp. 233-249.

Rosenber, R.S & Loomis J.B. (2001). Benefit transfer of outdoor recreation use values: A technical document supporting the Forest Service Strategy Plan (2000 revision). RMRS-GTR-72, Fort Collins, CO: U.S. *Department of Agriculture, Forest service, Rocky Mountain Research Station*

Vrhovcak, M.B.; Zeljiko, T. & Debrecin, N. (2005). External costs of electricity production: case study Croatia. *Energy Policy,* Vol. 33, No.11, pp. 1385-1395

Wagstaff, A. (1991). Health Care: QALYs and the equity-efficiency tradeoff. *Journal of Health Economics.* Vol. 10, pp. 21-41

Wang, H. & Mullahy, J. (2006). Willingness to pay for reducing fatal risk by improving air quality: a contingent valuation study in Chongqing, China. *Science of the Total Environment* Vol. 367, No.1, pp. 50-57

Wang, Y. & Zhang, Y-S. (2009). Air quality assessment by contingent valuation in Ji'nan, China. *Journal of Environmental Management,* Vol. 90, No. 2, pp. 1022-1029

Welsch, H. (2006). Environment and happiness: Valuation of air pollution using life satisfaction data. *Ecological Economics,* Vol. 58, No.4, pp. 801-813

Won Kim, C.; Phipps, T.T. & Anselin, L. (2003). Measuring the Benefits of Air quality improvement: a spatial hedonic approach. *Journal of Environmental Economics and Management,* Vol. 45, No. 1, pp. 24-39

Yoo, S.H.; Kwak, S.J. & Lee J.S. (2008). Using a choice experiment to measure the environmental costs of air pollution impacts in Seoul. *Journal of Environmental Management,* Vol.86, No. 1, pp. 308-318